The Skull of Quadruped and Bipedal Vertebrates

Comparative Anatomy and Posture of Animal and Human Set

coordinated by
Djillali Hadjouis

Volume 2

The Skull of Quadruped and Bipedal Vertebrates

Variations, Abnormalities and Joint Pathologies

Djillali Hadjouis

iSTE

WILEY

First published 2021 in Great Britain and the United States by ISTE Ltd and John Wiley & Sons, Inc.

ISTE Ltd
27-37 St George's Road
London SW19 4EU
UK

www.iste.co.uk

John Wiley & Sons, Inc.
111 River Street
Hoboken, NJ 07030
USA

www.wiley.com

Library of Congress Control Number: 2021931631

British Library Cataloguing-in-Publication Data
A CIP record for this book is available from the British Library
ISBN 978-1-78630-607-4

MIX
Paper from responsible sources
FSC® C013604

Contents

Introduction

One day I will go to live in Theory, because in Theory everything is successful.

Pierre Desproges
(1939–1988, author's translation)

If man cannot interpret in the future, and this is one of the hardest tests of his destiny, he can try to interpret in the past.

Albert Gaudry
(1827–1908, author's translation)

I.1. A series for the comparative anatomy of Mammals

Over the past 40 years, hundreds of thousands of human and animal vertebrate bones have passed through our hands. Skulls, horn cores, dentures, spines and limb bones, mainly from Plio-Pleistocene and Quaternary Mammals from North Africa, Switzerland and France, have been studied, compared, inventoried and classified. The same is true for human remains, but the interest focused on three chrono-cultural periods (the Upper Paleolithic, Neolithic and Middle Ages), two geographical regions (the Maghreb and the Paris Basin) with the support of two specialties (paleo-anthropology or bioanthropology and paleopathology), which have become inseparable for me because they are indispensable. The postural analysis that followed on Quaternary Mammals according to an architectural conception was beneficial later, when studies on the craniofacial postural dynamics of the genus *Homo* took over, according to the same concept but based on a standing posture. The upright position of the human body has changed centers of gravity and the distribution of gravitational forces is no longer balanced as was the case with quadrupeds. Today, arthrosis, arthritis, lower back pain, lumbosacral hinge anomalies and joint malformations are experienced daily by bipeds. It is in this spirit that this work was carried out, taking into account the evolution of the organizational plans of Mammals, their adaptation to the ground, the adequacy of the head posture in relation to different diets, dental articulations according to the fossil species of the genus *Homo* and the influence of an anomaly or a craniofacial pathology on not only the occlusion, but also on the infracranial regions.

The book *Atlas des maladies et traumatismes du monde médiéval et moderne : V*^{e} *siècle-XVII*^{e} *siècle*, published in French by ISTE Editions in 2018, was perhaps the trigger for the new set of books I am directing. Part of the "Biology" collection, entitled "Comparative Anatomy and Posture of Animals and Humans", the set brings together works dealing with all anatomical regions, either the holistic concept, integrating the architectural ensemble of the quadruped or bipedal vertebrate body and its postures, or a particular region (skull, teeth, limbs, spine, pelvis, etc.), or a particular joint (occipito-cervical hinge, shoulder joint, hip joint, knee joint, etc.).

The first book published in this series (June 2020) is by Dr. Cyrille Cazeau and is entitled *Foot Surgery Viewed Through the Prism of Comparative Anatomy: From Normal to Useful*. In addition to these, other works are to follow.

I.2. Introduction to the craniofacial and dental taxonomic terminology of vertebrates and dimorphism

The skull of Vertebrates and in particular that of Mammals envelops the encephalon while protecting it, while the face is made up of two articulated jaws. One, welded to the skull and immobile, is called the maxilla, and the other, the mandible, articulated to the skull through the temporo-mandibular joint, is mobile in order to produce all the necessary mechanical and dynamic movements (mastication, grip). Two fundamental developmental processes are to be considered in the formation of the skull. The ossification of these two ontogenic processes relies on the chondrocranium and desmocranium. In the first, ossification is achieved by a substitution process (enchondral ossification); in the second, membrane bones develop directly in the connective tissue (chondral ossification). The latter make up the majority of the bony scales of the cerebral skull and facial skull, while the bones formed from a cartilaginous outline are those located at the basi-cranial level (occipital, sphenoid, temporal with the petrous bone, ethmoid, hyoid bone (Kahle *et al.* 1980)). It will be seen later that in cranial malformations causing asymmetries, spheno-occipital synchondrosis (SOS) will be at the center of any ontogenic interpretation of an imbalance of this type, since its disjunction will be the cause or effect of the problem.

Studies on the skull, as on the rest of the skeleton, are nowadays analyzed from several angles, given the improvement of analytical tools, particularly through medical imaging. However, we must not neglect the fundamentals of anatomy and its twin sister, comparative anatomy, which in addition to the normal development of individuals, a great intraspecific variability is recorded in human species as in other Vertebrates, not to mention sexual dimorphism.

The paleontological analysis of the skull of *Hominidae* and quadruped Mammals involves a series of taxonomic, morphometric, geometric, cladistic, biomechanic and radiographic studies, grouped or isolated in order to recognize the different morphotypes, evolutionary mechanisms, ancestral origin of taxa, developmental stages or intra- and interspecific variability of each family, genus or species. The subspecies is itself considered as a geographic variety. All these studies are not necessarily carried out in the same research institution and may sometimes be unrelated to each other due to different objectives. Thus, research conducted in the medical world, particularly in dentofacial orthopedics, orthodontics, rheumatology or neurology, brings its share of knowledge to evolutionary sciences and the anatomy of fossil species, which sometimes prove to be very complementary. Throughout this work, a superficial or in-depth overview of these different research parameters will be provided, each time with examples supported by detailed illustrations.

Due to the radiation of certain Mammals, such as the *Bovidae* Family, the intra- and interspecific variation is great and manifests itself as much by polymorphism as by sexual dimorphism. Paradoxically, in some species of closely related Ungulates, such as small, large or medium-sized antelopes, difficulties appear during the taxonomic separation of the bony remains. This is particularly the case for certain dentures, cranial portions or limb bones belonging to gazelle species, but it is also the case for wildebeest (*Connochaetes*) females with male hartebeest (*Alcelaphus*). However, the most cited example in the zoological and paleontological literature is clearly that of the three or four genera (if we add *Bubalus*), *Bos*, *Bison* and *Syncerus* represented by the species of aurochs (*Bos primigenius*), bison (*Bison priscus*) and buffalo (*Syncerus*). Many authors have multiplied the methods of generic and specific distinction of these species with their imposing size, only the cranial remains, in this case the horn cores (orientation, curvature, flattening, etc.) and the metapodial segments provide fairly significant elements of diagnosis.

In caballine *Equidae* fossils, here again only the presence of canines in males sometimes allows a sexual diagnosis and this is not often the case because of a poor conservation of the symphysis elements in archeological excavations.

In Carnivores or in certain Great Apes, it is rightly the canine (or fang), clearly larger in males, which makes sexual determination possible, while the cranial vault of the majority of Primates is provided with a sagittal crest in males. Fossil *Hominidae*, on the other hand, represents a clear dimorphism whose cranial, pelvic and appendicular characters offer better elements of distinction. Another scenario is emerging for the current representatives of the human species where studies suggest that bony sexual dimorphism decreases greatly between men and women.

Cranial and in particular dental terminology has followed the methods of several authors: Gentry and Gentry (1978a) for Antelope jugal teeth, Heintz (1970) for *Cervidae* and *Bovidae*, Eisenmann (1980, 1981) for *Equidae* and Kurten for Carnivores (1963).

The different chapters discussed on the skull are based on current and fossil species of *Hominidae* and certain groups of four-legged Mammals. A large number of species studied by the author are based in particular on the cranial and dental forms of herbivorous and carnivorous species endemic to Eurasia, the Mediterranean Basin and Africa.

PART 1

The Skull of Fossil and Present-day Quadruped Vertebrates: Craniofacial Structure and Postural Balance

1

Proboscideans: The Mammoth (*Mammuthus primigenius*)

Super-Order *Proboscidea*.

Order *Elephantoidea*.

Family *Elephantidae*.

Genus *Mammuthus* Brooks, 1828.

Species *Mammuthus primigenius* Blumenbach, 1799 (woolly mammoth).

1.1. Chronological, geographical and morphological indications of the species

In the species of the *Elephantidae* Family, the orbits are very advanced on the antero-posterior axis of the skull and open in front of the jugal teeth; these are anatomical characteristics of animals without snouts. In general, in Mammals, the orbits open above the last molars (Lecointre and Le Guyader 2001).

The nose and upper lip are replaced by a flexible tube used for breathing, drinking, picking things up, etc. The lower canines are lost and the upper incisors are transformed into tusks that are continuously developing (Figure 1.1).

The species belongs to one of the three related genera regrouping mammoths and elephants (*Mammuthus*, *Elephas* and *Loxodonta*) included in the Family *Elephantidae* and classified in the Super-Order *Proboscidea*. These trunk bearers have provided no less than 170 fossil species, the oldest dating back 55 million years (Eocene). The first *Elephantidae* appeared in Africa 7 million years ago, the first mammoths 3–4 million years ago. Great confusion reigned for a long time among specialists for the classification of the mammoth and its origin. Today everyone is unanimous in classifying the first mammoths in Africa.

The arrival of mammoths in Europe and Asia took place around 2.6 million years ago. Three main Euro-Asian species followed one another. The southern mammoth (*Mammuthus meridionalis*), known in Saint-Vallier, Senèze and Chilhac, had a size exceeding 3 meters at the withers, large slightly curved tusks and jugal teeth whose hypsodontic character was still weak. The steppe mammoth (*Mammuthus trogontherii*), descendant of the previous one, was known in Sussenborn and Mosbach (Germany), in Abbeville and Saint-Acheul (France) and in Great Britain. With a height of more than 4 meters at the withers, this mammoth was considered the largest species in Europe. The woolly or Siberian mammoth (*Mammuthus primigenius*) is the best known. It appeared about 190,000 years ago and remains the most documented of all fossils due to the many discoveries of whole animals preserved in the permafrost (Guérin 1996a).

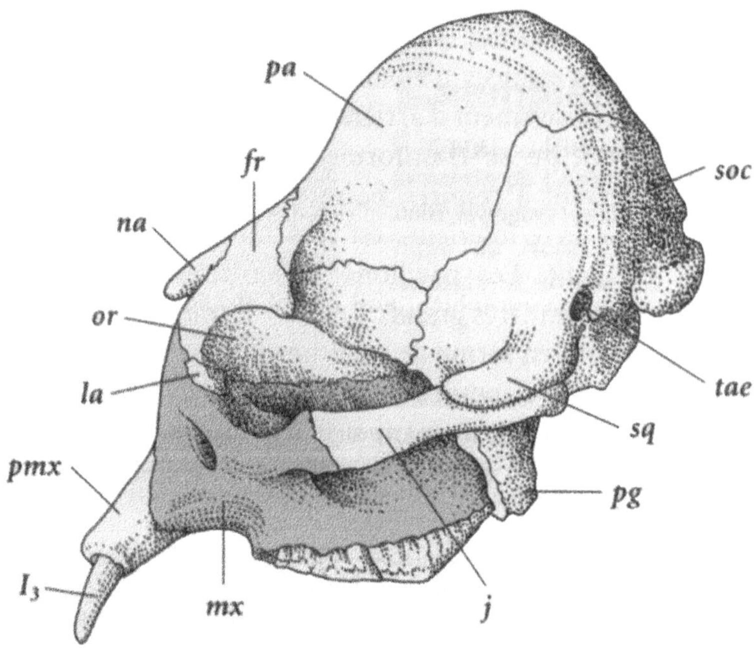

Figure 1.1. *Skull of a young present-day Asian elephant in left lateral view. Note the position of the orbits above the maxilla and not behind it (© Éditions Belin/Dominique Visset)*

1.2. Mammoth discoveries in Île-de-France

The *Fifth International Conference on Mammoths and their Families*, held in 2010 in Le Puy-en-Velay, brought together the discoveries of this trunk fossil in more than 20 countries of Eurasia, Africa and America. Even though the Île-de-France region was not represented at this meeting, there have been a large number of remains of woolly mammoths in the departments of Hauts-de-Seine, Seine-Saint-Denis, Val-de-Marne, Yvelines, Essonne, Seine-et-Marne and Val-d'Oise, most of which were cleared during the 19th and 20th centuries as a result of quarrying on the outskirts of Paris. The latter also had its share of discoveries, sometimes complete skeletons, such as that of the Montholon Square near the Montmartre cemetery. In the department of Val-de-Marne, several bony and dental remains of woolly mammoths (*M. primigenius*) have been found since the end of the 19th century on the banks of the Seine and its confluence with the Marne (Ardouin *et al.* 2009; Hadjouis 2020a). Thanks to the land development of the last 40 years carried out along the river banks, preventive archeological operations have brought to light new discoveries in well-dated biostratigraphic contexts.

Although the remains of ancient mammoths and elephants unearthed in this small department of south-eastern Paris have been numerous (remains preserved at the Muséum national d'Histoire naturelle, Musée de Saint-Maur-des-Fossés and Musée d'antiquités nationales), most of them were transported and deposited on the banks of the Seine or in the loop of the Marne (Le Perreux, Nogent-sur-Marne, Saint-Maur-des-Fossés, Créteil, Valenton, etc.) without great chronostratigraphic precision.

Appearing in Eastern Siberia around 800,000 years ago, the woolly mammoth's existence was known in Western Europe 200,000 years ago in chronocultural contexts of the Middle Paleolithic. With the exception of some spectacular specimens, it is mainly mandibles, cranial portions or isolated teeth that are found during archeological excavations or chance discoveries. Among the fossils recently dated by radiometric methods (uranium/thorium and carbon 14), two stand out: the young mammoth from Maisons-Alfort and the skull from Bonneuil-sur-Marne.

1.3. A young mammoth in Maisons-Alfort

The development operations of the Seine-Marne confluence, which began in the 1980s as part of the Seine-Amont project, have highlighted major sites such as Alfort 1 in Maisons-Alfort, located on the left bank of the Marne, a few hundred meters from its confluence with the Seine. One of the plots excavated in 1995 revealed a Neanderthal occupation of the Middle Paleolithic within the alluvial formations (Durbet *et al.* 1997). The association of identified animal species (aurochs, bison, red deer, Mosbach horse, wolf, mammoth) and the presence of a lithic industry shaped by Neanderthals allowed the reconstruction of paleoenvironments suggesting a cold and dry glacial climate. Several datings on fossil remains ranging from 160,000 to 200,000 years BCE give an image of a glacial environment in the Paris region inhabited by both humans and animals.

Usually, the attribution of Proboscidian species (*Elephas*, *Loxodonta* and *Mammuthus*) is based on complete third molars. It is indeed this last molar which delivers the greatest number of distinctive dental criteria (tooth shape, number of blade teeth and their index, enamel character and their plicature, sinus morphology, etc.). The milk tooth of the Maisons-Alfort baby mammoth was an incomplete lower third milky premolar (D3) with four blades (length: 36.2 mm; height: 44.6 mm).

1.4. A woolly mammoth skull in the reserves

The inventory of Vertebrate faunas carried out at the request of the director of the Val-de-Marne Departmental Archaeology Laboratory, Philippe Andrieux, in the 1990s included a fossil, which exceeded in volume and significance all other bone and dental remains. The skull without mandible of a woolly mammoth, perfectly preserved, had been waiting to be studied since 1923.

Without the short note *Les gros blocs quaternaires du port de Bonneuil* by Paul Lemoine and Teilhard de Chardin published in the journal *La Nature* in 1923, no one would know the exact origin of this Propboscidian. The authors describe the work carried out on Barbière Island, a commune of Bonneuil, located between the Marne and Morbras rivers, in order to dig the future coal port of Paris. The dredging of alluvium and blocks of local origin (coarse limestone with ceriths, Champigny limestone, Brie millstone and Fontainebleau sandstone) had yielded the fossil remains of *M. primigenius* and *Bos*. The woolly mammoth skull discovered during these dredging operations was not described in this note, but was obviously part of the port works.

It was in the 2000s that we studied it, starting with radiocarbon dating. C14 analyses of the radiocarbon laboratory of Villeurbanne gave dates around 45,000 years BCE, the Middle Paleolithic period. Although more recent than the dates of the Middle Paleolithic site of Alfort 1 (190,000 BC), which also yielded mammoth remains, the fact remains that the skull of Bonneuil figures as a spectacular piece within the Quaternary bestiary of Île-de-France. Although it shows traces of anthropic activity that suggest human intervention, there is nothing to distinguish between hunting and natural death. Similarly, the intervention on its carcass must have taken place in two stages after death.

Another individual was found in the same conditions in Bonneuil-sur-Marne, notably a tusk with traces of anthropogenic cutting.

1.5. A mammoth skull with removed tusks

As mentioned above, the Bonneuil-sur-Marne fossil was not found in well-documented stratigraphic levels or in a chronocultural context of the Paris Basin. However, its fortuitous discovery during the dredging of the coal port works in 1923 does not in any way detract from its exceptional character. The 45,000 year dating corresponds to a period of cold and dry climate in the Middle Paleolithic. Anthropic indications found on this skull without a mandible suggest hunting practices or the recovery of tusks from a dead animal.

The morphological criteria found on the maxillary teeth show the typical characters of the woolly mammoth (weak enamel with sinuous sinuses, high hypsodontia index and strongly folded enamel ribbons (Hadjouis 2016b)). The age given to this animal at the time of death is estimated to be between 22 and 34 years. It is based on criteria such as the partial wear of the molars, the clearing of their roots and the absence of wear of the last molar, some of whose distal blade teeth were still embedded in the alveolus. Three jugal teeth were still active on their maxillary medium: the second molars on both sides and the right third molar (Figure 1.2). The two anterior molars were in the process of being replaced.

The skull without its mandible, but especially without its tusks, is neither accidental nor a phenomenon of taphocenosis. The very careful cutting of the alveolar banks that surround the tusks testifies to their intentional recovery. Indeed, the posterior paired parts of the alveoli of the tusks were carefully and deeply cut out so as to be able to recover the ivory teeth in their totality.

This practice could only have been carried out by hunters or scavengers in two periods: either the individuals were contemporaries or they were post-Neanderthals. In any case, they recovered the tusks from the carcass of the slaughtered or naturally dead animal to transform ivory materials or for any kind of development.

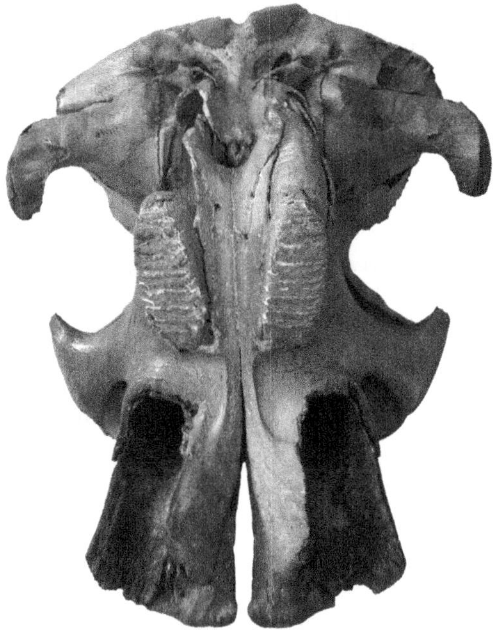

Figure 1.2. *Lower view of a mammoth skull (Mammuthus primigenius), found in the gravel of Bonneuil Port, showing the replacement of the jugal teeth (© Hadjouis)*

1.6. A particular tooth eruption

The most evolved Proboscidians of the genera *Elephas* and *Mammuthus* adopt/adopted a horizontal type of dental eruption of the jugal teeth, different from the other groups of the Mammalian class whose vertical eruption first concerns the deciduous teeth, replaced by the permanent teeth. In the woolly mammoth (*M. primigenius*), whose life expectancy was around 60 years, a second tooth eruption would not have been sufficient during its lifetime and all the teeth would have been worn down to the root, causing a final fall of the last stumps. Only the upper incisors called tusks saw two permanent and milky dentitions. To meet the biological longevity of the *Elephantidae*, a dental replacement cycle exists whose eruptive mechanism will be renewed six times. The mammoth's extinction coincided with the total stop of this cycle.

When the premolar and molar teeth were drilled, the mesial (anterior) blades of each jugal tooth appeared first, whose tooth wear was not balanced with the distal (posterior) blade teeth, because they were still embedded in the alveolus. This mechanism of pushing from the back to the front, especially in the case of M3s, caused the anterior teeth to fall out, regardless of the dental row (premolars or molars), as they erupted until the end of a cycle. In the end, it was the last molar (M3) that occupied the alveolar spaces of both jaws (Figure 1.3).

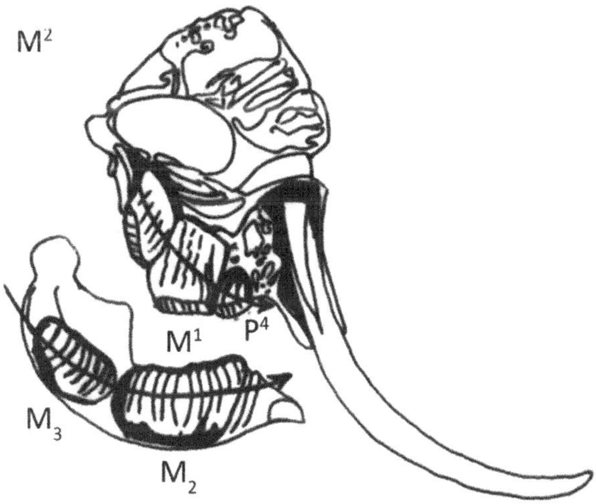

Figure 1.3. *Replacement of jugal teeth in the mammoth. Premolars and first molars were expelled from the back to the front, keeping only the M3 in each half jaw (© Anthony)*

2

Equidae

2.1. The horse (*Equus caballus*)

Order *Perissodactyla* Owen, 1848.

Family *Equidae* Gray, 1821.

Genus *Equus* Linnaeus, 1758.

Species *Equus caballus* (the horse).

2.1.1. *Chronological, geographical and morphological indications of species*

The current species belonging to the genus *Equus* Linnaeus *sensu lato* are divided into groups of horses (*E. przewalski*, *E. ferus* and *E. caballus*), zebras (*E. zebra, E. burchelli and E. grevyi*), hemiones (*E. hemionus* and *E. kiang*) and donkeys (*E. africanus* and *E. asinus*) (Groves 1986; Guérin 1994).

The dental formula is: 3/3 I, 0-1/0-1 C, 3/3-4 P, 3/3 M.

The domestic horse has short ears, a strong mane and long hair implanted along the entire length of the tail. These characteristics change according to the great equine diversity. Two types of domestic horses are recognizable: the light and fast "blood" horses and the heavy and powerful horses suitable for various service work. Among these varieties are: the Arabian thoroughbred, the English thoroughbred, the Frederiksborger horse, the Oldenburg horse, the Lipizzaner and breeds of ponies (Northern pony, Iceland pony and Shetland pony) (Huass and Petter 1900).

In addition to the dental characteristics, the base of horses' and donkeys' skulls can be distinguished by an occipital bone that protrudes over the dorsal face of the head in horses. The upper cranium forms a protruding external occipital protuberance. In donkeys, the more pronounced protrusion of the external occipital protuberance initiates an overjet curve on the nuchal face (Barone 1976). On the face, there are also specific peculiarities between these two species (the two frontal bones forming a flat surface between the orbits, with a large and strong zygomatic process in horses). The horse's parietal bone is characterized by a temporal line which joins its opposite by a sagittal ridge, non-existent in the young. The sagittal suture is fused late and the synostosis that begins from the rear at about 10 months is not completed until about 3 years of age (Barone 1976). Mandibular synchondrosis is fused from the age of 6 to 8 months.

However, the character that seems more important according to Lesbre, Petit and Barone (in Barone 1976) is that of the skull's angle of flexion on the face which is more pronounced in donkeys than in horses. The shortening of the spheno-basilar axis has resulted in greater craniofacial flexion in donkeys. The very much open basion-prosthion geometric triangle in horses reflects the large extension of the craniodental and craniospinal fields (Figure 2.1).

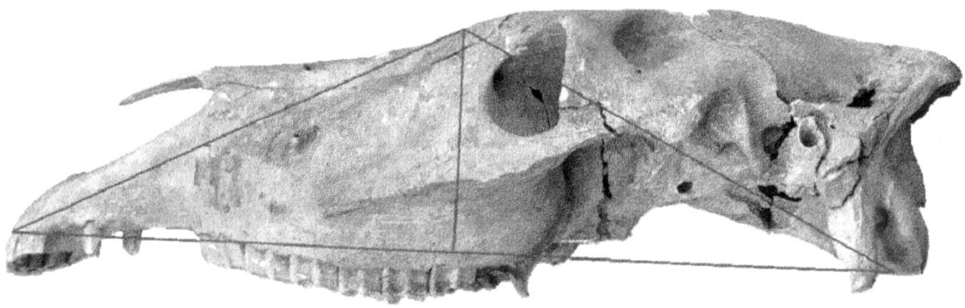

Figure 2.1. *Horse skull viewed from the left. The basion-prosthion geometrical triangle translates the extension of the craniodental and craniospinal fields. The perpendicular line of the triangle passes through M3 (© Hadjouis)*

The hybrid forms between the horse and the donkey give two infertile varieties: the mule, which is the product of the donkey and the mare, and the bardot, which is the product of the stallion and the donkey.

The Family *Equidae* made its first appearance in the North American Eocene. Two genera were recognized in Eurasia and Africa. The first, Christol's Hipparion, represents the last lineage of the tridactyl species. It arrived at the beginning of the Upper Miocene and survived in Africa until the Middle Pleistocene. Four species were defined: *Hipparion crissum* Déperet; *Hipparion fissurae* Crusafond and Sondaar; *Hipparion rocinantis* Hernandez Pacheco and *Hipparion crusafonti* Villalta. These were fairly large forms, with folded upper jugals, moderately hypsodont and isolated protocones (Eisenmann 1977; Eisenmann and Sondaar 1989; Guérin 1996b). The second, *Equus* Linnaeus, appeared in the Plio-Pleistocene with Stenon's horse. The latter *Equus stenonis* was characterized by primitive forms of the jugal teeth: jugal teeth superior to the narrow, simple and rounded parastyle and mesostyle, the protocone was short with a slightly elongated mesial lobe. In the lower ones, the medial lingual groove was narrow and angular. Compared to caballine-type horses, the limb bones had a less elongated metatarsal than the metacarpal, and the peri-articular muscular reliefs were more prominent (Prat 1980).

The genus *Equus* represents a large number of species that came to constitute the essence of the taxa evolving in Europe and France.

During their evolution, the *Equidae* have acquired new adaptations according to their biotopes (increase in size, specialization of teeth adapted to graminivorous regimes, monodactyl locomotor adaptation after being tridactyl, etc.). Among these major skeletal transformations, horses of the late Tertiary and early Quaternary adapted to open environments (prairie, steppe and savannah) with paleoenvironmental particularities. This is how we find running animals with slender limbs in open environments or stocky animals evolving in soft environments with snowy or sandy territories.

The vast sample that makes up the European forms of caballine horses (true quaternary horses) has allowed specialists to recognize about 10 species divided into three types. The first type includes stocky horses with small teeth and evolving in a temperate climate. The species that identify this type are *Equus taubachensis*, *Equus germanicus* and *Equus palustris*. The second type includes horses of medium to large size with a short and wide snout and frequenting cold fauna, represented by *Equus chosaricus* and *Equus arcelini*. The third type corresponds to medium to large animals, sometimes with shapes that have large teeth in both sizes. This type frequents cold or cool climates: species such as *Equus mosbachensis*, *Equus achenheimensis*, *Equus steinheimensis* and a small horse *Equus antunesi*, with a narrow snout and slender metapods. Subspecies close to this type have also been recognized in Tautavel (*E. mosbachensis tautavelensis*) and Lunel-Viel (*E. mosbachensis palustris*) (Prat 1968; Mourer-Chauviré 1972; Cregut 1980; Eisenmann *et al.* 1985; Eisenmann 1988, 1991).

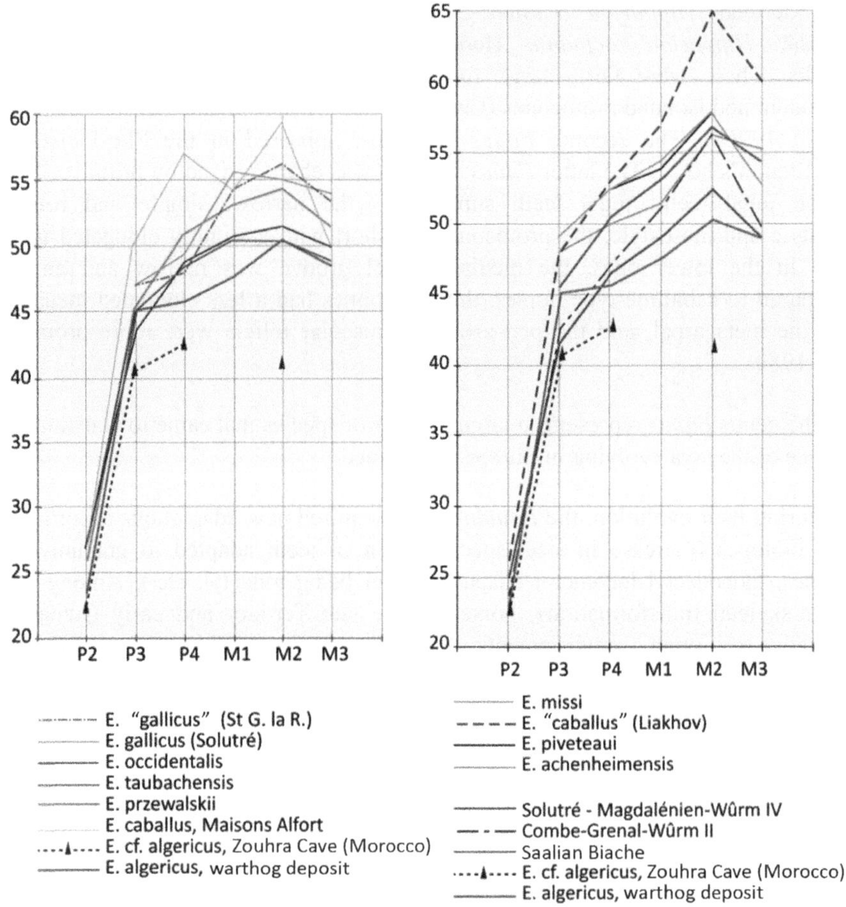

Figure 2.2. *Diagrams showing the comparison of the average protoconic indices of the definitive superior jugal teeth of several horses from Europe and Africa (© Hadjouis). For a color version of the figure, see www.iste.co.uk/hadjouis/skull.zip*

In his numerous works on *Equidae*, Eisenmann tried to give explanations on the cranial forms and those of the bones of the distal segments of fossil and current species. It is clear that on many interpretations, doubt persists in spite of everything, because morphological differences are also linked to genetic differences. Indeed, comparative analyses of the cranial remains of horses from the Middle Pleistocene of Europe (Mosbach, Caune de l'Arago and Lunel-Viel) show significant differences in the relative lengths of the snouts, which seemed to behave in the same way in other Pleistocene Caballines and behave in the same way in present-day ones. They could result from ecological adaptations (shortening of the snout in response to cooling) possibly leading to speciation (Eisenmann *et al.* 1985). Similarly, the European Quaternary horses, distinguished by the dimensions and proportions of their teeth and metacarpals, discriminate the large Antewürmian forms and the small ones from the Würm onwards. On the basis of the relative lengths of the protocones, this author defines three types with different sizes: type I with relatively long M1 and M2 protocones, adapted to a temperate climate; types II and III with short protocones, adapted to a cool or cold climate (Eisenmann 1991). The horse of the Saalian levels from Alfort 1 to Maisons-Alfort suggests that it belonged to the latter type (Hadjouis 1998a; Figure 2.2).

2.1.2. *A fossil horse in Africa: paleogeographic and biostratigraphic distributions*

Equus algericus (Bagtache *et al.* 1984) recognized in the Warthog Aterian deposit (Algiers, Algeria) (Bagtache and Hadjouis 1983; Bagtache *et al.* 1984; Hadjouis 1993, 2001, 2003a, 2003d; Hadjouis and Le Bihan 2014) was a caballine-type horse, measuring about 1.44 m at the withers, with stocky metapods and teeth with a wide and angular lingual groove and a very asymmetrical double loop typically caballine. It was found in the deposits of Ain Taya and Ain-Benian near Algiers, Oued Djebbana near Tébessa (Hadjouis 1993), Columnata in the Tiaret region (Chaid-Saoudi 1984), and Tit Mellil not far from Rabat (Hadjouis 1993), in Sidi Bou Knadel near Rabat, in El Mughara El Aliya near Tangier (Zouhri and Aouragh 1997) and in the Zouhra Cave in El Harhoura south of Rabat (Aouragh and Débénath 1999; Aouragh 2001); the latter are all dated to the Upper Pleistocene (Hadjouis 2001). Its reported presence at the Moroccan site to *Homo erectus* in Ain Maarouf near El Hajeb on the basis of a single tooth seems doubtful, especially since all the other *Equidae* remains have been attributed to the *Equus mauritanicus* (Geraads and Amani 1997).

This horse appeared during the Upper Pleistocene and probably corresponding culturally to the Moustero-Aterian periods (Middle North African Paleolithic –32,000 to –16,000). The Algerian horse (*E. algericus*) was certainly of Euro-Asian origin since no acceptable ancestor in Africa can be found at present; all the Plio-Pleistocene horses of this continent being Stenonian.

The dispersion of horses to the Maghreb during the Upper Pleistocene was from a center of Euro-Asian origin. This East-West migration by land that took place during the last glaciation is well known to authors. Several Eurasian species had to follow the migration corridor along the Gulf of Sidra. This was notably the case for the wild boar (*Sus scrofa*), the Algerian red deer (*Megaceroides algericus*), the red fox (*Vulpes vulpes*) or the rhinoceros (*Dicerhorinus hemithoecus*). It seems likely that *E. algericus* took part in these successive waves. Its passage through the

Moustero-Aterian material and Aterian sites thus seems quite logical. However, its absence in Epipaleolithic sites further complicates its status. While *E. algericus* did not appear in any of the Iberomaurusian and Capsian deposits in Algeria, its presence in this part of Africa would have been of short duration, which seems unlikely to us in view of the other Eurasian immigrants. Moreover, the Filfila deposit was not epipaleolithic (in the prehistoric sense), as has been pointed out by some, since no lithic industry or anthropic activity has been found there. It is a red clay fissure filling, corresponding rather to the periods of runoff resulting from the last glacial episodes (Ginsburg *et al.* 1968) and could be correlated to the Soltanian/Oulijian (isotopic stages 4 to 2) of the continental and marine chronologies of Morocco assuming that a Mediterranean/Atlantic correlation can be accepted (Texier *et al.* 1994). However, the presence of *Ammotragus lervia* at this site (if the determination is correct) gives this species an antiquity that has never before been raised. As for the presence of the Algerian horse in the Columnata Epipaleolithic (Iberomaurusian or Columnatian?), it rests on a distal end of a metacarpus, which in our opinion seems extremely limited without the association of sufficient dental and skeletal material to allow a distinction. However, the measurements nevertheless impose an approximation with the horses.

Compared to the *Equidae* of the Zouhra Cave in Morocco, *E. algericus* from the Warthog deposit was smaller than the Moroccan form but with similar protoconic indices. As for the European species, we had already mentioned in a previous work (Bagtache and Hadjouis 1983) the similarity in proportions and size of the Algerian horse with Jaurens' *Equus gallicus* (early recent Würm (Mourer-Chauviré 1980)), Solutrean (Prat 1968) and the horse from Torralba in Spain (Prat 1977). At the dental level, although the protocones are longer on P3/P4 than on M1/M2, its typological attribution (if this method were applied to vertebrates from Africa (Eisenmann 1991)) would be consistent with type 1, horses from temperate climates such as *E. gallicus* from Saint-Germain-de-la-Rivière, from Solutré and Combe-Grenal from Würm II (Guadelli 1987).

Concerning the horse (*E. caballus* or *ferus*), although its domestication dates back more than 3 millennia, known in Eastern Europe the species is determined with certainty in sites in France only in the Bronze Age. Sporadic remains of the wild form have been found in Neolithic sites of Val-de-Marne (Maisons-Alfort and Ivry-Grand-Ciel). The domestic species was attested to in the Gallic period by three forms of different sizes. The first one, with a slender shape, was a small horse with microdontia teeth. Its cranial, dental and skeletal characteristics bring it close to Iron Age horses with a height at the withers of 1.10 m. This size falls within the range of variation of average domestic horses of the Iron Age (between 1.10 m and 1.40 m) and therefore excludes the hybrid forms (bardot and mule), the pony and donkey. The second form had a height at the withers of 1.24 m also falling within this same variation, and the third, slightly larger, measured 1.39 m. These three forms of domestic horses nevertheless posed the problem of hybridization.

2.1.3. *The postural balance of* Equidae

The Family *Equidae* reached during its evolution the maximum of its specialization by the development of its third digit (also in the Rhinoceros and Tapirs), the second and fourth being vestigial. The last phalanx ending in a powerful hoof supporting a body with symmetrical postural balance in each leg (Figure 2.4). The cephalo-caudal axis shows a template whose withers and rump illustrate a masterful structure reminiscent of the suspension bridge (Figure 2.3). In fact, the fore and rear-end are in a front-to-rear axis at equidistant height. The spine, from the first four thoracic vertebrae to the caudal vertebrae, shows a curvature with a lower concavity. The cantilevered head is positioned high, due to a verticalized cervical spine, exceeding in height the withers by five cervical vertebrae. Because of this balance, the gravitational forces have developed the same robustness in the fore and hind limbs, including both sexes. In spite of a little apparent sexual dimorphism on the bones, only the absence of canines in the mare constitutes a difference.

Figure 2.3. *Posture of the horse's head and quadruped body. The Equidae are often used as an example to represent perfectly balanced posture and locomotion. The withers and the rump are located at the same height with a center of gravity positioned in the middle; the head is held high up by a verticalized neck that passes the withers by five cervical vertebrae (© Hadjouis and Le Bihan)*

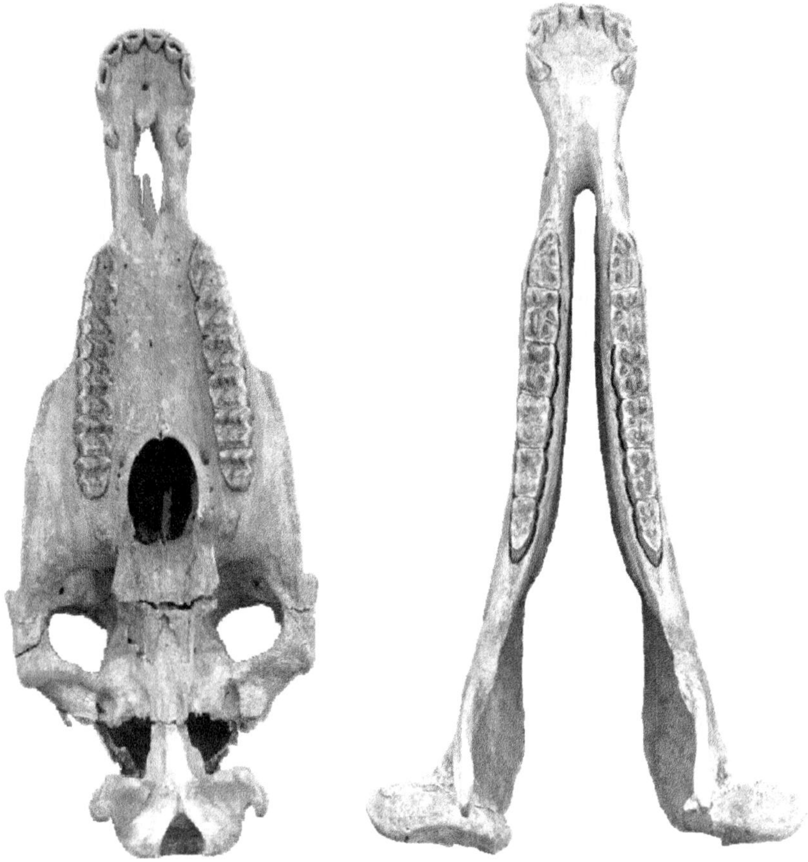

Figure 2.4. *Probable hybrid calvarium of a horse (Equus sp.) and mandible from above showing the symmetry of the paired dental and cranial parts (© Hadjouis)*

The wild and domestic forms, free of transport, draught and pack work, do not present any particular post-cranial asymmetries. Similarly, occlusion is perfectly balanced and only animals in captivity develop abnormalities or particular dental pathologies due to tooth wear against the bars of menageries.

2.1.4. *Joint pathologies in service horses*

Among domestic horses, with the exception of riding horses, many castrated forms that have allowed for increase in size have been used since Antiquity for multiple jobs. Those used for the transport of heavy loads or ploughing have developed severe arthropathies (osteoarthritis and/or arthritis) in the thoracolumbar spine and in the distal segments (carpals and metacarpals, tarsus and metatarsus). These osteo-articular manifestations have in some individuals stiffened entire regions (vertebral welds, carpo and tarso-metapodal welds, etc., Figures 2.5 and 2.6). Other bone diseases known to occur in horses include osteomalacia, developed especially in the brood mare during gestation, laminitis (inflammation of the flesh of the foot which deforms the hoof) most often resulting from forced labor on poor ground, recessed foot (tightening of the posterior parts of the foot) or bleeding (bruising produced by the tightening of the heels (Blanc 1983)).

Figure 2.5. *Inflammatory lesions involving osteoarthritis on a horse's lumbar spine, caused by heavy service work during the 19th century in Arcueil (draught and/or ploughing). Note the intervertebral osteophysical formation (© Hadjouis)*

Figure 2.6. *Inflammatory lesions involving osteoarthritis on the L4-L5 of horses, caused by heavy service work during the 19th century in Arcueil (draught and/or ploughing). Note the intervertebral osteophysical formation (© Hadjouis)*

The archeological sites of the Paris Basin have yielded skeletons of horses, donkeys and hybrid forms from all periods presenting important pathologies related to various jobs.

2.1.5. *Introduction to animal bone pathologies and zoonoses*

Since the Paleolithic, mosquitoes and animals have been in contact with humans through which the majority of infectious or parasitic agents have had as breeding grounds different species, whose great contamination revealed its first symptoms during the Neolithic period.

Some of these infectious diseases, especially those of epidemic origin such as the plague (*Yersinia pestis*) or parasitic diseases such as malaria (*Plasmodium falciparum*), are systematically researched by laboratory teams, and particularly the skeletons that are part of the burial complexes that are currently being excavated. Because zoonoses (diseases transmitted by animals), vectors of parasitic illnesses developed after the sedentarization and breeding of animals since Neolithic times, are becoming more and more significant by promoting the emergence of pathologies carried by domesticated species: mange, roundworms, ringworm in dogs, bovine tuberculosis, measles, chickenpox, distomatosis in cattle, Malta fever, small fluke in goats and sheep, tapeworm in pigs, influenza in migratory birds and plague in rodents (Biraben 1995). While the epidemic spread has occupied these territories for several centuries, we must not neglect thalassemia, also known in this same Mediterranean region, particularly in Greece since the Bronze Age. For Biraben (1976), the existence of the plague before its penetration into Europe can be traced back to the 2nd century BCE. Its traces were found in the Mediterranean all along the route corresponding to the late Phoenician expansion, and also to the Roman conquest. The example of tuberculosis is well known in Egypt through DNA, radiography and macroscopy, while elsewhere it remains unknown as for other infections, because it is not systematically researched due to the lack of adequate programs.

2.1.6. *The horse's status over the centuries*

The horse's status during Antiquity was very high, to the point where the binomial of the species *Equus* (name of the genus) and *caballus* (name of the species, designating gelding or pack horse) took on a double meaning. The qualities and virtues of the horse were praised in the Middle Ages, considering the species as faithful to its master and ardent in combat. The horse was also used as a service animal, in daily life (an animal for transport, draught and pack animal) (Duchet Suchaux and Pastoureau 2002). The place occupied by the horse during the Middle Ages was linked to the development of equestrian enterprises.

In the foreword by Assia Djebar of the *académie française*, written for Salah Guemriche's dictionary of French words of Arabic origin (2007), she says about horses:

> I give myself the free pleasure of parading in front of you, precisely in equestrian art, the 9 terms of which at least 6 are a summary of the different breeds of horses. Imagine, dear reader, that you dreamed of having, supreme luxury, in your stable, 6 exceptional horses: one day you would ride the *chestnut* – which, according to its Arabic root, is distinguished by the reddish-fawn color of its coat; on the second day, your choice would be a *zain* horse – whose coat is of a single color without white fur; on the third day, your preference would be for a *strawberry roan* horse – with a grayish, but mottled color; on the fourth day, you would be proud of your *aubin,* which is a small horse from Ireland … Then perhaps your *gelding* – which is neutered – would stay in the stable; and on the sixth day, you would reserve the *jennet*, a small, fast horse, for yourself. (author's translation)

The texts on equestrian art in the Arab-Muslim world are legion. In the only text that was miraculously saved from the fire that targeted the destruction of the Arabic manuscripts stored in the Spanish monastery of El Escorial in 1671, Ibn al-Awam, a famous Sevillian Arab agronomist of the 12th century, describes his agronomical work. Dozens of pages are devoted to the horse alone, considered an animal with unequalled virtues. He describes, in addition to the way of treating the sick animal, the ways of riding a horse, with or without weapons (Clément-Mullet 2000).

In the 19th century, the unpublished sum of documents on horse riding and hippology gathered by General Daumas (1803–1871) in the book *The Horses of the Sahara* illustrates the relations that the Arabs had with the horse, its breeding, its education, the care to be given to it and the thousand ways of using it (Pouillon 2008).

2.2. The donkey (*Equus asinus*)

Order *Perissodactyla* Owen, 1848.

Family *Equidae* Gray, 1821.

Genus *Equus* Linnaeus, 1758.

Species *Equus asinus* Linnaeus, 1758 (African wild donkey).

2.2.1. *Chronological, geographical and morphological indications of species*

Two subspecies of wild donkey are known in Africa and particularly in North Africa: the Nubian wild ass, *Equus asinus africanus* (or *Equus africanus africanus*), native to Nubia, Sudan and Eritrea, and the Somali wild ass, *Equus asinus somaliensis* (or *somalicus*). The Somali wild ass is said to be the origin of all domestic forms; it is recognizable by the transverse stripes on its legs. However, while the interspecific distinction is biologically clear-cut, stating that donkeys belong to a species distinct from hemiones, zebras and horses, the complexity seems great when distinguishing between current domesticated donkeys and their ancestors (Groves 1966; Eisenmann 1995). According to Groves (1966, 1986), neither the Nubian nor the Somali wild ass appear to be good candidates for the role of ancestors to the domesticated donkeys and do not constitute homogenous groups, judging by their outward appearance. In general, zoologists distinguish wild donkeys by, among other things, their colors: the wild Nubian form is grayish or gray, without dark markings on the limbs, the Somali form is reddish fawn, the shoulder stripe is poorly defined or absent, and the limbs are marked with dark transverse stripes (Dorst and Dandelot 1976).

To this must be added a possible third subspecies of wild donkey, recognized on the rock walls of Algeria (Camps 1984).

The domesticated donkey is smaller than the horse, has long ears, a heavy, upright head, a poorly developed mane, and the hair is formed only by a small cluster extremity (Barone 1976; Figure 2.7). The frontal bone is distinguished from that of the horse by its proportionally larger zygomatic process, and the supraorbital edge is also more prominent above the orbit than in the horse. The parietal bone, which has the same disposition as the horse, is traversed by a temporal line interrupted by a more or less wide groove that separates it from the external sagittal ridge. This characteristic does not always exist in the horse (Barone 1976).

An asinian species (*Equus melkiensis*) was first described in 1983 by Bagtache *et al.* at the Aterian Warthog site in Algiers (Algeria), later found at other sites in Morocco. Melki's donkey is the largest and the most robust of the asinian species in Africa: its size varies between 1.35 m and 1.40 m at the withers. The morphological characteristics of the teeth and their occlusal pattern are similar to those of the asinian species, while one of the dental characteristics (typical stenonian pattern and relatively symmetrical double-loop) is found in species other than donkeys, particularly in hemiones.

Figure 2.7. *Head and body quadruped posture of the African wild ass (© Hadjouis and Le Bihan)*

It has been found in the typical Warthog deposit, in the deposit of karst cracks in the Filfila massif, in the Aïn-Benian cave (Bagtache and Hadjouis 1983; Bagtache *et al.* 1984), in Columnata (Chaïd-Saoudi 1984) and in Morocco (Sidi Bou Knadel and El Mughara El Aliya (Zouhri and Aouragh 1997)). For Eisenmann (1995), some of Tighennif's metapodes could be related to this same species, making *E. melkiensis* the oldest asinian in Africa, known only in North Africa. This hypothesis seems difficult to accept when we know that all dental material is attributable to the zebra *E. mauritanicus* alone. Furthermore, we remain attached to the idea that phylogenetic relations could exist between *E. melkiensis* and *Equus tabeti* of the Ain Hanech (Hadjouis 2001). Recent research on donkeys in Africa suggests that genomic analyses are indispensable for a better ancestral identification of the fossilized Asinian taxa as well as for their descendants (Youcef 2018, 2020).

2.2.2. *The status of the donkey over the centuries*

According to popular tradition, if the donkey is humble, the horse is proud, because of its humility, hard work, obedience and patience. But most often it has been designated, first by the Fathers of the medieval church, as a symbol of ignorance, stupidity, stubbornness, laziness and lechery, to such an extent that the image has been assigned to man (Duchet Suchaux and Pastoureau 2002).

3

Bovidae

3.1. Aurochs (*Bos primigenius*)

Order *Artiodactyla*.

Family *Bovidae* Gray, 1821.

Genus *Bos* Linnaeus, 1758.

Species *Bos primigenius* Bojanus, 1827 (aurochs).

3.1.1. *Chronological, geographical and morphological indications of species*

Large herbivorous ungulates with oversized horns, aurochs and bison, species of the largest Family of Mammals (*Bovidae*) long cohabited together in the open spaces of the cold and temperate periods of the Quaternary. One (the aurochs) was a ubiquitous animal and adapted to various climates, both glacial and during periods of interglacial and temperate warming, while the other acclimatized mainly to the steppes, these immense meadows of herbaceous plants, an environment also conducive to the development of the woolly mammoth.

They were characterized by hypsodont jugal teeth with selenodontal occlusal surfaces. The head had frontal appendages called horn cores. These were permanent and covered with a horny case that was not preserved in archeological deposits. In general, taxonomic identification was made on the basis of the shapes of the horn cores (orientation, curvature, spiraling, flattening, etc.) (Figure 3.1) and the general shape of the skull, as well as on the metapods. The generic distinction on the teeth remains very difficult, despite several attempts by paleontologists and archeozoologists. In the domestic cattle as in the fossil, the two mandibles never fused together and remain mobile on each other throughout life (Barone 1976).

The dental formula is: 0/3 I, 0/1 C, 3/3 Pm, 3/3 M.

Figure 3.1. *Frontal and 3/4 right views of a cranial portion of aurochs (Bos primigenius), with its horn cores from Aflou (Algeria) (© Hadjouis)*

Aurochs or great wild cattle were native to Europe and Western Asia, and their southern distribution during the Quaternary spread as far as North Africa. It was seen from the beginning of the Middle Pleistocene in Europe (Ilford, Swanscombe, Steinheim, Lunel Viel, Tautavel, Terra Amata and Le Lazaret). In the absence of complete bone remains, cranial and dental elements, it was easily confused with the remains of steppe bison, hence the name "Great *Bovidae*". In North Africa, remains were abundant from the Upper Pleistocene onward and, apart from the cranial remains, which were confused with the buffalo (genus *Syncerus*). On the other hand, the paleoecological interest in these Great *Bovidae* is of prime importance, as shown by the adaptive characteristics provided by the distal segments of the locomotor skeleton.

In the Paris Basin, and more precisely south of Paris, the remains of aurochs (and probably bison) have been found in sites dating from the Middle Paleolithic (Maisons-Alfort 1) to the Final Neolithic (Maisons-Alfort 2, Ivry-Grand-Ciel and Port de Bonneuil). Still present in the Neolithic, the remains of aurochs were still in a minority compared to red deer and wild boar. The new rules of herd management and hunting controls applied during the Holocene are more selective in relation to certain wild species, this is the case for the red deer and wild boar, which are much more numerous in terms of numbers.

3.1.2. *Cattle (*Bos taurus*)*

The Early Neolithic communities that settled in the territory of southern Paris took advantage of the choices offered by natural habitats. Indeed, sites were found in varied environments (valley bottoms, the Seine-Marne confluence, the banks of the Seine in the flood zone, the Villejuif and Thiais plateaus) and the strategies for the exploitation and subsistence of livestock and its derivatives were managed in the same way as in the rest of temperate Europe of the 6th–5th millennium. In the classically adopted trilogy (cattle/goats and sheep/pigs), everywhere the breeding of cattle did not include other species of medium or small size. The same management was practiced in the habitats of the Val-de-Marne. Examples of domestic cattle in the Early Neolithic period (Villeneuve-Saint-Germain) were represented in the joint development zone of Rungis les Antes and the joint development zone of Alfort 2.

In the Middle Neolithic, here too, the systems of game exploitation were based on the feeding behaviors established in the west and south of the Paris Basin. Domestic cattle were still widely represented in the Paris Basin sites. In the Val-de-Marne, the sites on the plain and along the banks of the Seine of the Cerny and Chasséen cultures, present in Maisons-Alfort and Ivry-Grand-Ciel, show that more than 35% of domestic species were selected for meat production. Peaks in mortality concerning this animal during these periods were distributed between 2 and 4 years, oriented toward a meat economy. There were also older cattle, kept for milk production and/or draught needs.

The Final Neolithic and Early and Middle Bronze are poorly represented in archeozoological terms in the south of Paris, with the exception of Alfort 2, and it is from the Final Bronze Age onward that the bone remains of domestic species were exploitable for the protohistoric periods. In the period of the Final Bronze Age and then in the First Iron Age (800 BCE), the exploitation of large animals such as cattle took second place to goats and sheep. After having been the main species in the food management of domestic livestock in previous periods, cattle were found here, used as a complementary resource to the breeding of goats and sheep. This characteristic, well-illustrated in the occupations of the Marne loop, was also found in the plateau sites or on their outskirts. In the Second Iron Age (La Tène), the orientations of cattle breeding underwent once again new modifications. Cattle once again became the main domestic species, with goats and sheep providing food supplements.

3.1.3. *The status of cattle over the centuries*

Although it is widely consumed, the status of cattle is somewhat displaced because of the new zootechnical exploitations imposed on them with the Roman conquest. In fact, the castration techniques used on the bull, in order to increase the anima's size, made it lose its symbolic image. In addition, the economic and food yields (consumption, draught and transport animals) of this great bovine are well deserved. During the Middle Ages, in spite of very large exploitation, cattle were also considered with many virtues.

The symbolic image of cattle in Antiquity was antinomic, sometimes they were seen as castrated bulls, an animal that had lost its ardor, sometimes represented as symbols of domestic and patrimonial wealth. The animal was not free but subject to a master to whom it must be submissive, dependent and obedient (Duchet Suchaux and Pastoureau 2002).

3.2. The bison (*Bison priscus*): chronological, geographical and morphological indications of the species

> Family *Bovidae* Gray, 1821.
>
> Genus *Bison* Smith, 1827.
>
> Species *Bison priscus* Bojanus, 1827 (steppe bison).

Much larger and stronger than aurochs, the steppe bison originated from Europe and North Asia. Its horn cores of single-curved horns were shorter than those of aurochs, but problems of generic distinction remain due to the absence of the latter. In spite of attempts to use dental methods to separate the two taxa, which remain unsuccessful, only postcranial elements such as cervical vertebrae and metapods have succeeded in this exercise (Boessneck *et al.* 1963; Martin 1987; Serre and Hadjouis 1989, etc.).

The oldest specimens date from the Upper Pliocene of East Asia. Several Eurasian and North American subspecies of the steppe bison have been identified (*Bison priscus voigtstedtensis* Fischer, *Bison priscus gigas* Flerov, *Bison priscus priscus* Bojanus and *Bison priscus mediator* Hilzheimer).

Other forms also existed with much discussed origins such as: *Bison schoetensacki* Freudenberg and *Bison bonasus* Linnaeus.

3.3. The buffalo (*Syncerus antiquus*)

Family *Bovidae* Gray, 1821.

Sub-Family *Bovinae* Gray, 1821.

Genus *Syncerus* Hodgson, 1847.

Species *Syncerus antiquus* Duvernoy, 1851 (the ancient buffalo).

3.3.1. *Chronological, geographical and morphological indications of the current* Syncerus *and* Bubalus *buffaloes*

African buffaloes can be distinguished from Asian buffaloes by a series of characteristics, particularly in the skull: short and broad face, absence of fusion between the vomer and palate, short horns inserted immediately behind the orbits and molarization of the mandibular P4 (fusion of metaconid and paraconid of P4).

The genus *Syncerus* is based on the species-type *Syncerus caffer* (Sparmann 1779), the cape buffalo, which has a very wide geographical distribution south of the Sahara and could be considered monospecific. A mammal with the appearance of a large cattle species, the African buffalo, because of its great variability, has several forms, the most extreme type being the African forest buffalo (*Syncerus nanus*) inhabiting the forests of West and Central Africa. According to Freshkop (1955), their hybridization in the wild is not excluded.

Asian buffaloes are represented by two species: *Bubalus arnee* and *Bubalus depressicornis*. *Bubalus arnee* has four subspecies that are distinguished by the size and color of the skin (Epstein 1971): *B. arnee arnee* Kerr; *B. arnee pulvus* Blanford; *B. arnee bosei* Lydekker; *B. arnee mindorensis* Hende, and the Tamarao. The latter is the smallest domesticated form, its height at the withers is 1 m to 1 m 20 cm. *Bubalus arnee*, the wild water buffalo or Asian buffalo, lives in India, Sri Lanka, Indochina, Sumatra, Java and Borneo.

Bubalus depressicornis (the Anoa) is the smallest form of all the current wild *Bovinae* living in the Sulawesi Mountains. Its height at the withers is between 0.60 m and 1 m, and its horns are rather reminiscent of antelope horns. Groves (1981) proposes that Anoa be considered a subgenus of *Bubalus* along with the species *B. depressicornis* and *Bubalus quarlesi*.

The domestication of Asian buffaloes is connected especially in all of South Asia, Indonesia and the Philippines. Buffaloes were also found in Europe as early as the Middle Ages in southern Italy, Hungary, the Balkans and southern Russia (Huass and Petter 1900).

While no African buffaloes have been domesticated, in Egypt four types of buffaloes have been domesticated in the Nile Valley; one type in Upper Egypt and three in Lower Egypt (Mason 1969). The different buffaloes of Lower Egypt are identified according to their characteristics and distribution: Minufi buffalo in the province of Minoufia, Beheri (sea buffalo) in the province of Beheira and Baladi buffalo. The type of Es Said (Upper Egypt) is the Saidi buffalo.

3.3.2. *Chronological, geographical and morphological indications of fossil species*

The massive nature and above all the heaviness of the fossil forms made their cranial and above all skeletal identification as regards the generic separation with *Bos* easier, suggesting paleoenvironmental reconstructions (see below).

Fossil remains of the genus *Syncerus* are represented by the species *S. caffer*, of which an almost complete skeleton is known from the Omo in the Kibish formation (Leakey 1969). Other remains attributed to this same species have been found in Melkbos in South Africa and Zululand (Hendey 1968; Gentry and Gentry 1978a). A second species, *Syncerus acoelotus* Gentry, is represented in Oldoway material (Bed II to IV) (Gentry and Gentry 1978a).

In North Africa, the species-type *Bubalus antiquus* was described by Duvernoy in 1851 in reports by the Académie des Sciences and was based, in addition to the description of the Asian water buffalo, on a portion of a skull found in excavations during the foundation of a mill on the banks of the Oued Bou Sellam near Sétif. Philippe Thomas described other remains based on discoveries in the Rhummel gorges in Aïn Smara and in Oued Seguen. Pomel (1883) devoted a monograph in the geological map of Algeria, based on an almost complete skeleton found in a clay-marl deposit in Djelfa (Figures 3.6 and 3.7).

Figure 3.2. *Posture of the head and quadruped body of the ancient buffalo (Syncerus antiquus), 3/4 view (© Hadjouis and Le Bihan)*

The taxonomic complexity of the fossil species of African buffaloes, maintained by their similarity to the bovine species of the Eurasian genus *Bos*, has made their classification difficult for more than a hundred years and it has been modified several times. A reminder of the data concerning the genus *Bubalus* and the species type *B. antiquus* has been repeatedly proposed (Hadjouis 2015). Indeed, the preserved skeleton parts (teeth and bones) of the two main large bovines, buffalo and aurochs in North Africa and bison and aurochs in Eurasia, are rather easy to determine (Serre and Hadjouis 1989). However, this confusion has led to revisions with the creation of synonymies that are not justified. However, as early as 1949 the British paleontologist Miss Bate had recognized and reported that *B. antiquus* was a *Syncerus*, followed by Gentry with some doubt. Later Hadjouis (1985a, 1985b) confirmed this by insisting on kinship relations, from the Upper Pleistocene in particular, with *Syncerus* and not with *Pelorovis*, as Gentry wrongly pointed out (Gentry and Gentry 1978a).

Syncerus antiquus (synonymies *Pelorovis antiquus*, *B. antiquus* and *Homoïoceras antiquus*) was represented in Northern Africa from the beginning of the Upper Pleistocene (Figure 3.2). Its remains found in Algeria were from Oued Bou Sellam (Thomas 1876), Djelfa (Pomel 1894), Relilaï (Vaufrey 1955), Pointe Pescade and Aïn Benian (Arambourg 1931), Tamar Hat (Arambourg *et al.* 1934), Medjez (Bouchud 1975), Amekni (Bouchud 1975), at the Warthog deposit (Hadjouis 1985a, 1985b, 2003a, 2003d; Hadjouis and Le Bihan 2014), at the Ouled Nail mountains (discovery by Hadj Bel Abbès, Delloula and Tahiri in 1985, according to Hachid

(1992)). In Morocco, it was found in Tit-Mellil (Arambourg 1938), Bou Knadel (Michel 1990) and El Herroura (Aouraghe 2001). In Tunisia, the skull of *Buffelus palaeindicus* Falconer, found in the region of Bizerte (Falconer 1924) was none other than the ancient buffalo (Bate 1951; Gentry and Gentry 1978a).

Figure 3.3. *Frontal view of a cranial portion of a fossil buffalo (Pelorovis howelli), with its horn cores from the Lower Pleistocene in El Kherba (Algeria) (© Sahnouni)*

We will not return to the presence of the genus *Pelorovis* in North Africa and its taxonomic confusion with *Syncerus* in the Upper Pleistocene sites. Its presence in the lower Pleistocene levels of the Ain Hanech in Algeria (*Pelorovis howelli*; Hadjouis and Sahnouni 2006) is discussed elsewhere by Hadjouis and Le Bihan (2014) and especially developed by Hadjouis (2017) (Figures 3.3 and 3.4).

Figure 3.4. *Head and body quadruped posture of Howell's buffalo (Pelorovis howelli), 3/4 view (© Hadjouis and Le Bihan)*

Elsewhere in Africa, the oldest mention of *Pelorovis* concerns *Pelorovis oldowayensis* (Reck 1928). This species common in Olduvai was found in the middle and upper levels of Bed II. Kanjera and the Shungura (Omo) formation also delivered remains of *P. oldowayensis*.

Pelorovis antiquus (repositioned in the genus *Syncerus*) has been found in East Africa in Broken Hill (Zambia) and in Bed IV in Olduvay (Gentry and Gentry 1978a, 1978b). In South Africa, its remains are numerous in Cornelia, Elandsfontein and Florisbad (Cooke 1974; Hendey 1974; Vrba 1976).

Other species of fossil buffalo attributed to the genus *Pelorovis* such as *P. nilsonni* (Lonnberg 1933) from Kenya and *P. baini* (Seeley 1891) from South Africa are not specified in relation to the genus *Syncerus*.

The range of ancient buffaloes extends to the Middle East, reported in Lebanon by Bate (Garrod and Bate 1937).

Asian fossil buffaloes come from the *Proamphibos* Pilgrim and *Hemibos* Falconer lineage.

It is in this context of study and revision of Quaternary fossil collections that the subspecies *Syncerus antiquus complexus* (Hadjouis 2002) (Figure 3.5), recognized in Algeria, was located in an Aterian deposit (Hadjouis 2002, 2019a). This complex ancient buffalo is rich in an extremely rare osteological collection, in comparison with the remains of the same group, recognized in the Maghreb sites. Indeed, the morphometric characteristics of this new taxon not only differ from other known buffalo species by size, but the individuals are also represented by juvenile and adult populations of both sexes, which has allowed biometric, paleoenvironmental, sexual dimorphism and, finally, generic and specific dimorphism studies between *Bos*, *Bison* and *Syncerus*.

Work carried out on the morphology of the neck vertebrae and on the distal segments, in particular the metacarpals of these three bovines, has shown that the lengthening of the cervical vertebrae, a sign of adaptation to running, shows not only a relationship with the morphology of the metacarpals and the position of the body's center of gravity, but also with how the head is held. In fact, in aurochs, the raised head position characterized by leaf-eater type feeding is related to an elongation of the metacarpals and distal ends having a width proportional to the length of the bone. Such a disposition suggests locomotion on hard ground (Hadjouis 1985a, 1985b, 2006).

Figure 3.5. *Head and quadruped body posture of the ancient buffalo subspecies (Syncerus antiquus complex). In the example of the buffalo, and the muskox, the center of gravity is placed at the front of the body, making withers higher than the rump, and the neck and head are positioned lower than the withers (© Hadjouis and Le Bihan)*

This positioning is different in buffaloes and fossil and current bison, due to a low head position, related to a feeding mode of grazer and grass shearer types. In these ungulates, the metacarpals are relatively short with wide distal extremities predisposing them to locomotion on soft ground (Hadjouis 1985a, 1985b, 2006; Serre and Hadjouis 1989).

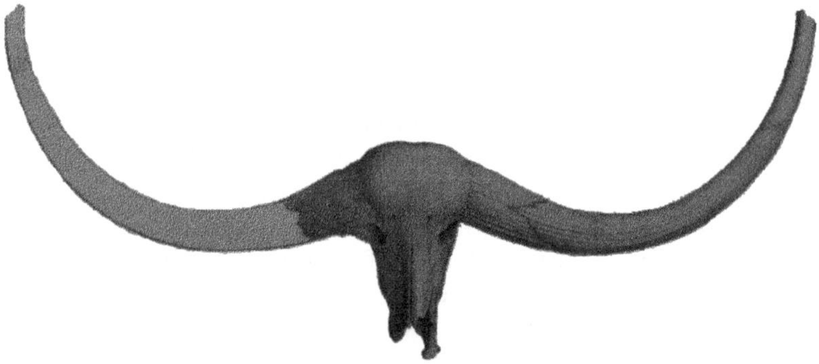

Figure 3.6. *Front view of the ancient Algerian buffalo (Syncerus antiquus) from Djelfa showing the wingspan, orientation and curvature of the horn cores (from Pomel 1883)*

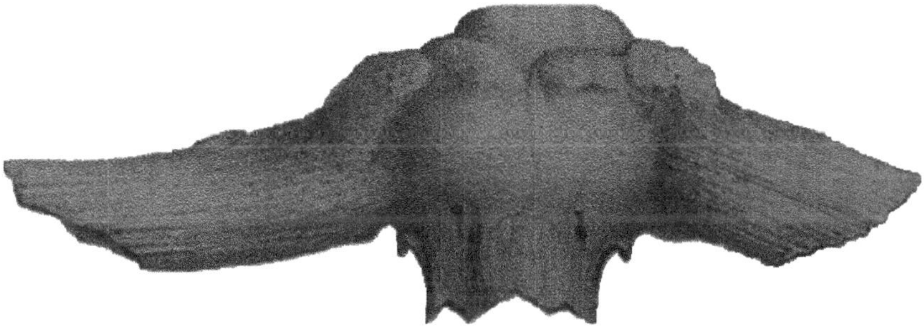

Figure 3.7. *Front view of an ancient buffalo skull (Syncerus antiquus) from Algeria (© P. Thomas)*

3.3.3. Bos/Syncerus *dental distinction criteria*

3.3.3.1. *Distinction of the upper teeth*

Among the many fossil remains from North Africa and the Mediterranean Basin, morphometric studies of the jugal teeth have shown clear differences between the genera *Bos* and *Syncerus* (and buffalo *sensu largo*). The morphological features of the occlusal surface and the coronal shaft retained for geometric distinction defined on the upper and lower molars are: styles and pillars of the outer wall, shape of the entostyle, presence or absence of interfossettes (enamel islands) and shape of the upper M3 metastyle.

3.3.3.2. *Styles and pillars of the outer wall*

On the vestibular surface of the upper molars of aurochs, the parastyles, mesostyles and metastyles were slightly angular, those of ancient buffalo were rounded. The foundations of the outer wall were remarkable for their protrusion and their more or less sharp angulation in aurochs, they are less prominent in buffalo.

In general, the vestibular faces between styles and pillars are rectilinear in *Bos,* they are more or less rounded in *Syncerus*.

3.3.3.3. *Shape of the entostyle*

This shape is very characteristic and distinguishes the two genera. In *Bos*, the molar series carries an entostyle whose morphology and implantation are variable, regardless of tooth erosion. In lingual view, the entostyle has three distinct morphologies: a rounded entostyle placed in the middle of the lobes, a pointed and very protruding entostyle and an entostyle fused to the distal lobe.

In *Syncerus*, the morphology of the entostyle is presented lingually as follows: it originates at the base of the distal lobe, develops along the shaft and spreads out on either side of the two lobes. In occlusal view, the entostyle appears in two different forms: split entostyle and flattened entostyle.

3.3.3.4. *Presence or absence of interfossettes*

The majority of the upper molars of *Bovidae* that we have studied show interfossettes in both genera; however, it should be noted that their presence was more frequent in ancient buffalo. Given the absence of enamel islands on juvenile jugal teeth or teeth barely damaged by tooth erosion, we performed two horizontal sections on the shaft of each of the two teeth of *Bos primigenius* and *S. antiquus*.

The two sections performed correspond to two new stages of wear that completely change the occlusal pattern of the tooth:

– on a straight M2 of *B. primigenius*: in stage 1, which corresponds to the initial stage, the tooth was slightly worn away. It was noticeable that the distal edges of the protocone and mesial edges of the hypocone penetrated deeply, intercalating between pre- and post-fossette. In stage 2, the well-worn tooth already took on a rectangular shape with an increase in vestibulo-lingual diameter. The distal edges of the protocone and mesial edges of the hypocone now formed an island, or even several central islands. The entostyle, which at the previous stage was isolated, united with the protocone and the hypocone. In stage 3, the tooth was very worn. It had a robust appearance, the pillars and styles were completely rounded, the fossettes were simplified and the enamel islands thinned. The three stages of wear had very different occlusal patterns. Naturally, it was at stage 3 that the tooth was easily confused with that of an antique buffalo, square-looking and robust at the same stage of wear;

– on an M3 of *S. antiquus*: stage 1 took up the same occlusal surface as the one sectioned on the aurochs' tooth. In stage 2, the occlusal surface changed completely. The walls of the protocone and hypocone withdrew and joined with the entostyle, thus forming central islands. At stage 3, a clear simplification of all characters appeared. In total, two characters allowed us to differentiate between the upper molars of the aurochs and those of the ancient buffalo according to the different stages of wear: flattening of the entostyle against the two lobes in *S. antiquus*, and the pillars of the outer wall, less prominent and less convex in this same species.

3.3.3.5. *Shape of the upper M3 metastyle*

The M3 is the only tooth among the isolated superior jugal teeth that allows generic discrimination. The latter differs from M1 and M2 by its generally trapezoidal shape, where the obliquity to the distal lobe is very strong in relation to the mesial lobe. But it is above all the morphology and development of the metastyle that are significantly distinct. In *B. primigenius*, the metatstyle was angular, sometimes rounded, but never in the form of a blade tooth protruding over the entire height of the tooth. The sometimes lamellar character of the base of the metatstyle fades toward the top. In *S. antiquus*, the metatstyle was thin and at its base it was constituted by a very developed flattened blade tooth which curved progressively toward the inner side of the tooth. This blade tooth was narrowed at the top.

3.3.3.6. *Distinction of the lower teeth*

A first *Bos/Syncerus* distinction was attempted by Thomas (1977), whose most significant criteria are presented in Table 3.1.

Bos primigenius	*Syncerus antiquus*
High horizontal ramus	Low ramus height
Thin mandible	Strong thickness of the mandible
Strong curvature of the mandibular fissure	Weak curvature of the fissure
Short interalveolar portion	Extended interalveolar portion
Clearly streamlined tooth pillars	Barely streamlined pillars
Presence of a second column	Absence of a second column at M3
Interlobal at M3	

Table 3.1. *Distinguishing criteria for Bos primigenius and Syncerus antiquus*

3.3.3.7. *Molarization process*

Concerning P3 of the genus *Bos*, the protoconid is not rounded and the metaconid-entoconid junction is primed, leaving an enamel island; this is none other than the remains of the entoconid and metaconid walls. Paraconid and metaconid are separated. The absence of soldering of the paraconid and metaconid suggests that these premolars belong to *B. primigenius*. In *S. antiquus*, the premolars were already molarized at this stage.

Among the isolated permanent premolar teeth, P4 from *Bovidae* and more generally from Ruminants has been the subject of much discussion. Arambourg (1947) described the structure of P4 in some large African *Bovidae*, notably the genera *Menelikia*, *Alcelaphus*, *Kobus* and *Hippotragus*, and wrongly acknowledged that *Bovidae* and *Hippotraginae* have “primitive type” premolars. Heintz (1971) in his work on Cervidae and some *Bovidae* shows precisely a molarization of the advanced P4 of *Addax nasomaculatus* (*Hippotraginae*). Other authors have shown that P4 molarization (metaconid-entoconid and metaconid-paraconid fusion) was complete in *P. oldowayensis* (Gentry and Gentry 1978a) and *S. antiquus* (Hadjouis 1985a), as well as in other *Bovidae* such as *Cephalophini*, *Reduncini* and *Alcelaphini*.

3.3.4. *Postural balance and paleoecology of* Bovidae

The horizontal cephalo-caudal posture of a large number of quadruped Vertebrates is linked to gravitational functionalities distributing loads according to balanced forces between the front and hindquarters. These structural combinations between craniofacial morphology, the frontal appendages with continuous growth, the morphology of the neck vertebrae, feeding mode and the locomotor dynamics, especially of some ungulates, have nevertheless had significant effects on the centers of gravity. In aurochs, the center of gravity was in the middle of the body, predisposing the withers and rump in the same plane (Figure 3.8). This arrangement is somewhat different in buffalo and bison, however, due to an earlier migration of the center of gravity to the front of the body. However, the sitting position of the front end is better due to this repositioning of the withers in relation to the hindquarters (Hadjouis 2006b).

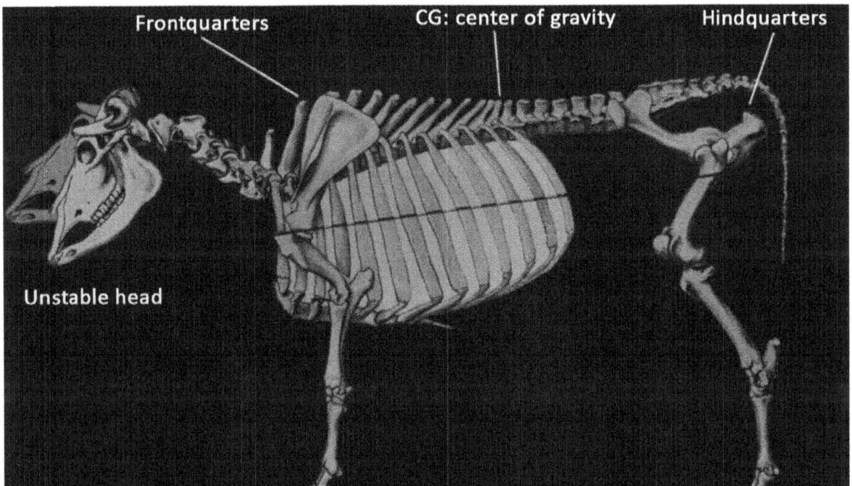

Figure 3.8. *Head posture and quadruped body of aurochs (Bos primigenius). Note the postural balance between the withers and the hindquarters, with the center of gravity positioned in the center of the body. The top of the head did not exceed the withers (according to Ghetie and Mateesco (1977) modified)*

In addition, in buffalo and muskoxen, the front end is increased by an elevation of the withers due to the excessive height development of the spinous processes of the first thoracic vertebrae. In the latter, the thrust of the spinous processes is such that the structural configuration of these *Bovidae* gives them the appearance of quadruped humpbacked animals with a low profile at the hindquarters. On the other hand, in cattle and buffaloes, they have a head that does not protrude beyond the withers. Only the gigantism of the bony horn cores of aurochs exceeded in height, sometimes starting from the pedicle, the horizontal cephalocaudal line.

The occlusion of these Vertebrates is balanced and does not present any asymmetrical manifestations. In the same way the craniofacial dynamics are in harmony. These two connected factors are in reality free of any postero-cervical mechanical stress, unlike bipeds. Moreover, the arrangement of several centers of gravity, one in the middle of the body balancing the front and hindquarters, the others functioning on the left and right sides, has considerably reduced possible postural asymmetries. Only the effects of domestication, those of captive animals, work and service animals will develop osteo-articular anomalies or degenerative malformations such as arthrosis or arthritis.

3.3.5. *Polymorphism and dimorphism in* Bovidae

For a large number of species, the intraspecific variation is significant because of strong polymorphism and/or dimorphism, especially sexual.

The presence in these *Bovidae* of certain preserved elements of the appendicular skeleton such as metacarpals and metatarsals allows them to be used thanks to certain methods which have the advantage of distinguishing by the same biometric methods, the genders and separating the domain of males from that of females. The technique used is that borrowed from G.G. Simpson (1941), described as "ratio-diagrams" and modified by Eisenmann (1979). Graphical data allow comparisons of size and shape to be represented. The method consists of converting the averages of the measurements into decimal logarithms, the logarithms for a selected species (*Bos taurus*) are taken as a reference base and represented by the line 0. The logarithms of the other measurements of fossil species are then plotted, in order to see the difference between the reference logarithms and those of the species to be compared.

To better grasp the paleogeographic and biostratigraphic implications of *B. primigenius*, the aurochs of certain paleolithic and paleontological sites of the Maghreb (Filfila, Warthogs and Tamar Hat) were first compared to the aurochs of the Middle Pleistocene of Europe whose deposits did not possess another *Bovini* such as the Bison. This is notably the case of Lunel-Viel in the Hérault (France), then to the Middle Paleolithic deposits of Jebel Qafzeh in Israel and to more recent deposits such as the Swiss Neolithic site of Bürgachisse-Sud.

These comparisons made it possible to distinguish the different forms of these *Bovini.* Indeed, one will notice, for example, that the aurochs of the Middle Pleistocene of Europe were more slender and more robust than those represented in the Aterian Warthog deposit (Algeria); however, the male and female distal extremities (joint and supra-articular distal diameters) were larger in the Algerian *Bos*. Having the same shapes and proportions, the size of the aurochs of Filfila and Tamar Hat was also much more slender than the aurochs of the Warthog deposit. Generally speaking, the *Bos/Syncerus* distinction can be summarized as follows:

– *Bos* metapods are always more slender and have more developed cross-sectional diameters than *Syncerus* metapods;

– a greater overall length is often accompanied by a strong development of diaphyseal cross-sectional diameters;

– in both *Bos* and *Syncerus*, the metatarsals are always more developed than the metacarpals;

– in *B. primigenius*, the supra-articular transverse diameters of the metacarpals were always smaller than the articular transverse diameters, whereas in *Syncerus*, they are slightly smaller;

– the proximal transverse diameters of the metacarpals and metatarsals were always smaller than the distal transverse diameters for both aurochs and the ancient buffalo. In modern cattle, the opposite is true;

– the distal diameters of the articular shanks are more developed on *Bos* male and female metatarsals.

Concerning sexual dimorphism in *B. primigenius*, the diagrams of the reports of average dimensions are significantly reliable and can be summarized as follows:

– males had a greater overall metapodal length than females. This was not always the case for the metatarsals, which sometimes had almost equal length between the two sexes;

– proximal transverse diameters always outweighed proximal anteroposterior diameters in *B. primigenius*;

– the diaphyseal diameters of the metacarpals were much more robust in males;

– males had very developed distal diameters compared to females;

– the anterior-posterior diaphyseal diameter is one of the most reliable characteristics. In females, the metatarsals always had a diaphyseal anterior-posterior diameter that outweighed the diaphyseal transverse diameter, a trait not found in the metatarsals of male aurochs; this explains the gracefulness and slenderness of female individuals due to the flattening of the transverse diameter of the diaphysis.

This same generic and sexual distinction was also visible on one of the tarsal bones like the talus. The *Bovini* from these sites have also, thanks to this bone, allowed the separation of aurochs and ancient buffalo as well as the establishment of criteria for sexual diagnosis (Hadjouis 1985a, 2003).

3.3.6. *Osteoarticular abnormalities and bone pathologies in* Bovidae

Large *Bovidae* have several types of pathologies, but those that are most often noted are first of all related to services (draught, transport and ploughing animals) in domesticated forms since the Neolithic period and whose arthropathies (arthrosis and arthritis) are recognized. Other pathologies are congenital, metabolic and especially infectious.

Concerning the domestic forms, the same type of osteo-articular afflictions found in domestic equine forms (horse, donkey and pony) is found in cattle. Indeed, it is the distal segments of the limbs that are affected first, developing primary osteoarthritis lesions in the carpo-metacarpal and metacarpo-phalangeal joints of the foreleg and the tarsal and metatarsal joints of the hind leg. Most often, these articular areas are ankylosed by great osteophytotic development whose osteoplastic anarchy has deformed the articular and periarticular parts.

In the same way, gravitational overload on the skeleton of the limbs, due to this arduous work, gave rise to a modification of the distal segments, particularly at the distal extremities of the metapodes. In the latter (metacarpals and metatarsals), the supra-articular part is concave and its articular part is even more developed in width. The articular eminences are also modified, with the laterals no longer symmetrical to the medians. This asymmetry sometimes creates joint eminences that are exaggeratedly outwardly or inwardly oriented. Some shapes have crossed anterior limbs in the manner of a human *genu valgum*. Indeed, Ghetie and Mateesco (1977) found that Neolithic cattle used for pulling resulted in a greater opening of the ante-brachio-carpo-metacarpal angle of up to 196° and a deviation toward the median axis of the bone rays of 162° (Figure 3.9). This exaggerated opening of the wrist angle works in the same way when both stationary and in propulsion.

The inflammatory mode of infectious diseases concerning the animal species found in archeological excavations is not often published and cases found do not easily find comparative examples. The best-known cases in the bone field concern the oral sphere and teeth (osteomyelitis, actinomycosis, infected abscess, etc.), syphilis in prehistoric Bears, osteomyelitis, rinderpest, cancers and post-traumatic infections.

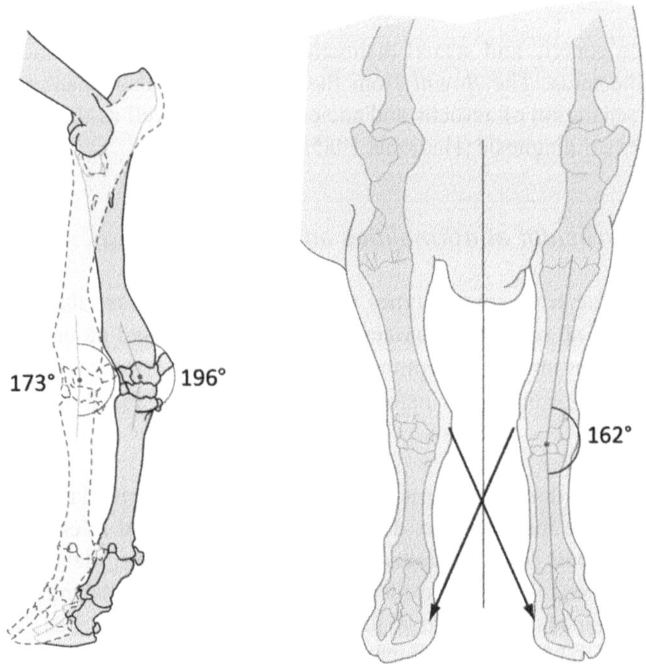

Figure 3.9. *Anterior limb of domestic cattle showing the effects of pulling during the Final Neolithic (large opening of the ante-brachio-carpal-metacarpal angle up to 196° and deviation toward the median axis of the bony rays of 162°) (from Ghetie and Mateesco (1977) modified). For a color version of the figure, see www.iste.co.uk/hadjouis/skull.zip*

The Aterian Warthog deposit yielded a frontal part with its bony horn cores belonging to the holotype of *Syncerus antiquus complexus* Hadjouis. While the right portion of the ankle is preserved just after the pedicle, the one on the left presents a portion atrophied by an infectious phenomenon or post-traumatic inflammation. Frontal sinus and cornual processes are one and the same, whose bone remodeling has caused the two antero-posterior surfaces of the ankle to flatten and develop significant osteosclerosis related to reactive subperiosteal osteophytosis at the pedicle/ankle junction. This invasive remodeling has caused the sinus cavities to disappear from the cornual process. Neighborhood infection is found in the left part of the frontal planum. The latter is scattered, since the intercornual protuberance, of thick and grouped layers of periosteal apposition (Figure 3.10).

The differential diagnosis is an infection (osteitis) in response to post-traumatic inflammation, or a primary cancerous tumor.

Figure 3.10. *Frontal with the start of the horn cores of the subspecies type of Syncerus antiquus complexus Hadjouis (ancient fossil buffalo from Algeria) from above. The start of the left horn core from the pedicle is atrophied by a probable infectious form (© Hadjouis)*

3.4. The common eland (*Taurotragus oryx*)

Family *Bovidae* Gray, 1821.

Tribe *Tragelaphini*.

Genus *Taurotragus Wagner*, 1855.

Species *Taurotragus oryx* Pallas, 1776 (common eland).

3.4.1. *Chronological, geographical and morphological indications of species*

The *Tragelaphini* are the largest species of African antelopes, represented even today by two species: the common eland (*Taurotragus oryx*) and the giant eland (*Taurotragus derbianus*). Although these remains are sporadic in the Quaternary sites of Africa, it remains that in their presence the doubt is excluded, especially if the preserved remains are represented by portions of horn cores, cranial portions, dental series or limb bones, some of which are characteristic (humerus, radius, metacarpus and metatarsus). This is the case of the remains of the common eland in the Upper Pleistocene deposits of North Africa, following the example of the Algerian Warthog deposit.

The two current species are also related to a third species, *Tragelaphus strepsiceros* (Pallas), the greater kudu (Figure 3.11). While the latter has horns of similar shape, their impressive length, divergence and two to three spiral turns make all the difference. But these examples are aggravated when they are present in prehistoric sites in Africa. Divergences have appeared in the taxonomic designation of these remains in the Maghreb sites. These are the genera *Taurotragus* (Pomel 1894; Hadjouis 1985a, 1985b, 1990, 2003a, 2003d, 2005; Hadjouis and Le Bihan 2014) or *Tragelaphus*, such as the example of *Tragelaphus marocanus* and *Tragelaphus algericus* (Geraads 1981) (Figures 3.12 and 3.13).

Figure 3.11. *Posture of the head and quadruped body of the great kudu (Tragelaphus strepsiceros), close to the common eland. The withers are higher than the hindquarters and the neck carries the head well above the frontquarters (© Hadjouis and Le Bihan)*

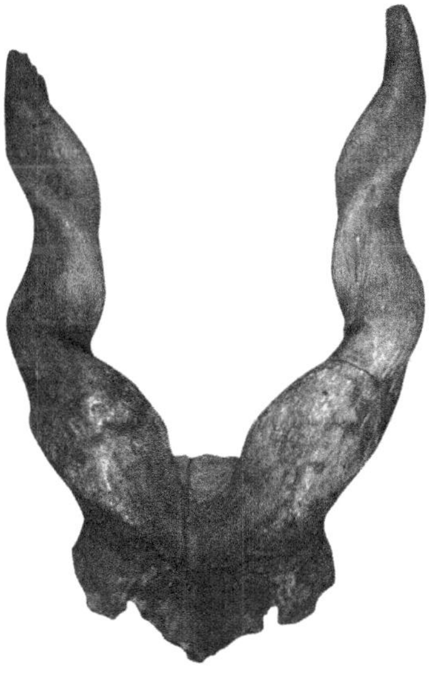

Figure 3.12. *Front view of a cranial portion of a fossil tragelaphine (Tragelaphus marocanus), with its bony double-twisted horn cores from the Schneider quarry (Morocco) (courtesy of P. Taquet, © Hadjouis)*

Figure 3.13. *Postural balance of the head of the common eland (Taurotragus oryx). Despite imposing spiral horns on a massive head, the latter is carried on a neck above the withers (© Hadjouis and Le Bihan)*

Taurotragus oryx, identified in Algeria at the Warthog site, is considered to be the most documented in terms of the number of remains in these regions of Northern Africa (five individuals, four males and one female, about 100 remains, 26 of which can be determined by specific classification) (Hadjouis 1985, 1990, 2005; Hadjouis and Le Bihan 2014; Hadjouis and Roubet 2020). The dental remains found are determinable at the supraspecific classification (brachyodontic dentition, streamlining, deep fossettes, absence of entostyle, highly developed M3 metastyle, molarization of the lower P4, presence of a short interlobar column on the lower molars, etc.). As for the skeleton of the limbs, it also suggests features that resemble this large ungulate (obliquity of the base of the internal trochlea of the humerus and notch of the medial lip, absence of an angular fossette at the level of the medial throat of the proximal joint of the radius, diaphysis of the radius, wavy and not rectilinear as in the giant eland or in other *Tragelaphini*, very slender metapods with a general length greater than that of the present eland but shorter than that of the giant eland, flattened base of the diaphysis of the metapods, distal transverse diameters of the distal extremity of the larger metapods, anterior edge of the metacarpus with a rounded and not convex section, faceted metatarsals for the large cuneiform and the very open cubo-navicular, etc.) (Hadjouis 2001).

3.4.2. *Posture and locomotor adaptation*

Figure 3.14 shows the segment diagram of the limbs of the Great *Tragelaphini*, such as *T. oryx*, *T. derbianus*, *Tragelaphus imberbis* and *T. strepsiceros*. The average length of the males is given according to the values of the Great *Tragelaphini* established by Obussier and Ernst (1977).

The stature of *T. oryx* (common eland) and *T. derbianus* (giant eland) has more slender frontquarters than the considered species of the genus *Tragelaphus* and particularly *T. strepsiceros* . However, the metacarpus of the great kudu is the slenderest of all. The humerus of *T. oryx* is always higher than the radius, whereas in *T. strepsiceros* the opposite is true. The ratio radius × 100/humerus is 96.74 in common eland, and 107.63 in great kudu (Hadjouis 1985a).

The hindquarters show the same differences in the same sense, but this time the metatarsals are almost equal between the species *T. oryx* and *T. derbianus* and *T. strepsiceros* . The femur of *T. oryx* is larger than the tibia, whereas the opposite occurs in *T. strepsiceros* . The tibia × 100 to femur ratio is 96.37 in *T. oryx* and 111.15 in *T. strepsiceros*. The calculation of this ratio gives us information on the adaptation of the hind limb to jumping or running. *Tragelaphini* seem particularly adapted to running and the fossil metapods discovered in Pleistocene sites in the Maghreb confirm this (Hadjouis 1985a).

Some authors noted that adaptive traits are strongly biotope-related. Antelopes with high withers, as in this case, occupy open habitats.

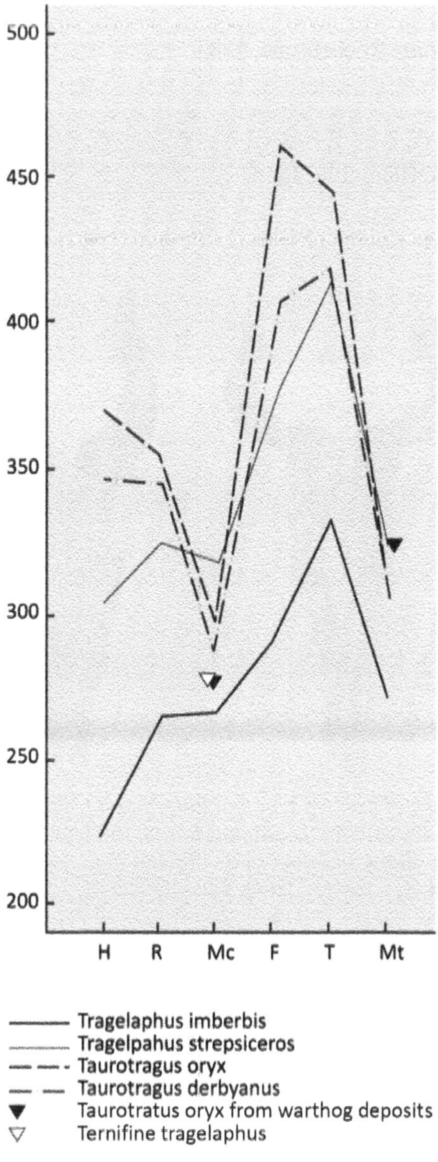

Figure 3.14. *Diagram of the segments of the limbs of the Great Tragelaphini (Taurotragus oryx, Taurotragus derbianus, Tragelaphus imberbis and Tragelaphus strepsiceros) and comparison with the Tragelaphini fossils from the Warthog and Ternifine sites*

3.5. The hartebeest (*Alcelaphus buselaphus*)

Family *Bovidae* Gray, 1821.

Subfamily *Alcelaphinae* Rochebrune, 1883.

Tribe *Alcelaphini*.

Genus *Alcelaphus* de Blainville, 1816.

Species *Alcelaphus buselaphus* Pallas, 1766 (hartebeest).

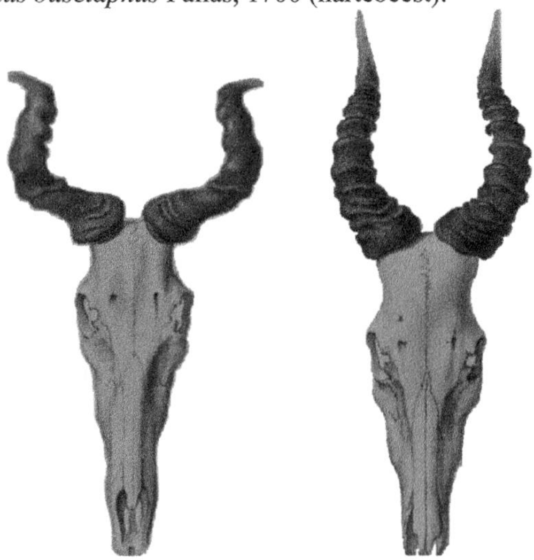

Figure 3.15. *Skulls of the two types of hartebeest (Alcelaphus buselaphus). The type on the left has U-shaped horns, the type on the right has horns in the shape of horizontal curls (source: gallica.bnf.fr/BNF)*

3.5.1. *Chronological, geographical and morphological indications of species*

Alcelaphinae are medium to large African antelopes with horns of diverse morphology, present in both sexes. The face is long and narrow with a frontal region extended toward the top, which is taken over by the pedicule, giving the face a particular morphology unlike any other group within the family. The horns have variable shapes, but two main types seem to emerge: a V or U-shaped type seen from the front and an S in profile (western group from Senegal to Chad, Cameroon and Central African Republic) (Figure 3.15), another type in the shape of a horizontal curl seen from the front (Blue Nile, eastern Ethiopia and Somalia, from Kenya to central Tanzania) (Dorst and Dandelot 1976).

Teeth are hypsodont, the pillars of the external and internal wall of the upper and lower jugal teeth are weakly convex, entostyle and entostilyde are absent on the molars and P4 is molarized.

In the Quaternary sites, its presence was clearly noted as early as the Upper Pleistocene and became very abundant in the Epipaleolithic sites, particularly in the Capsian deposits of Algeria and Tunisia.

3.5.2. *Postural balance*

The antelope has a peculiar profile, with a sloping back showing notably withers clearly higher than the rump (Figure 3.16). These characteristics follow, as in the *Tragelaphini*, with withers higher than the hindquarters. The hindquarters are adapted to a jumping and/or running locomotion in open savannah environments.

Figure 3.16. *Posture of the head and quadruped body of the hartebeest (Alcelaphus buselaphus). The center of gravity is displaced at the frontquarters due to the withers being higher than the hindquarters, the neck and head are positioned above the withers (© Hadjouis and Le Bihan)*

Male metacarpals and metatarsals from archeological sites can easily be confused with the same bones belonging to female or slender individuals of other antelopes of similar size, such as blue wildebeest (*Connochaetes taurinus*), roan antelope (*Hippotragus equinus*), scimitar oryx (*Oryx algazel*) or addax (*Addax nasomaculatus*).

3.6. Gazelles (*Gazella*)

Family *Bovidae* Gray, 1821.

Tribe *Antilopini*.

Genus *Gazella* de Blainville, 1816.

3.6.1. *Chronological, geographical and morphological indications of species*

The *Antilopini* tribe includes many forms of gazelles today, which are made up of representatives ranging from small to medium size. The distinctiveness of the *Antilopini* and particularly that of the genus *Gazella* is the least known due to the low variability of the anatomical characters. Representatives of these forms are currently known from Africa, Asia and Arabia (Dorst and Dandelot 1976).

There are two groups of gazelles on the African continent. The first is represented by species from the Paleoarctic domain and extends only as far south as the Sahara. These species are *Gazella dorcas* Linnaeus (the dorcas gazelle), *Gazella cuvieri* Ogilby (Cuvier's gazelle), *Gazella leptoceros* Cuvier (rhim gazelle), *Gazella rufina* Thomas and *Gazella isabella* Gray (the ariel gazelle).

The second group, Ethiopian, is represented by species currently distributed in West and East Africa. These are: *Gazella rufifrons* Gray (red-fronted gazelle) and *Gazella dama* (Pallas) (dama gazelle).

Gazella thomsoni Günther and *Gazella granti* Brooke both live partly in East Africa. The large Grant's gazelle with its massive horns has a large intraspecific variation.

The biometric characteristics of these *Bovidae* are based on the shape, curvature, divergence and spiraling of the horns, sometimes very imposing, the dental segments of the premolar and molar series (brachyodontics and hypsodontics) and the length of the limb bones.

In fossil species, the upper and lower teeth seem to lack reliable criteria. However, two characteristics seem to indicate their specific determinations: the calculation of the index of the lengths of the dental segments (PM/M) and the morphology of the upper M3.

These criteria have been applied to some fossil gazelles from Africa, in this case *Gazella atlantica* Bourguinat and *G. cuvieri* Ogilbi. The teeth of *G. atlantica* (Atlantic gazelle) are brachyodont and robust, but can easily be confused with the teeth of *G. cuvieri*, which are indistinguishable from the metastyle of M3. The ribs of the outer wall are developed and the lobes are slightly rounded. The mesostyle of P4 of the Atlas gazelle may be decisive, given its position in the middle of the tooth. In Cuvier's gazelle, it is positioned anteriorly, like that of P2 and P3 (Hadjouis 1985a).

The M3 metastyle of *G. atlantica* has a slight expansion visible from top to bottom, which distinguishes it from other gazelles. Concerning the lower teeth, the morphology of the posterior lobe of M3 also distinguishes certain species of gazelles.

Calculation of the ratio of the premolar segment to that of the molars makes it possible to biometrically distinguish several gazelle species, notably *G. atlantica*, *Gazella tingitana* and *G. cuvieri*, which have developed premolar lengths and the same dimensions adapted to a grazer-type diet (Gentry 1964; Hadjouis 1985a).

3.6.2. *Postural balance*

Small to medium-sized antelopes and gazelles are generally carried by a slender, graceful body with long, slender limbs that allow them not only to run fast, but also to jump up to 9 m in length and up to 3 m in height (Delachaux and Nestlé 1976; Figure 3.17). Their weight varies from 20 kg (*G. dorcas*) to more than 80 kg (*Aepyceros melampus*). Unlike some large antelopes, gazelles have balanced front and hindquarters. The withers and hindquarters are on the same plane, the head raised and carried by an upright cervical spine, in the manner of *Equidae*.

Figure 3.17. *Posture of the head and quadruped body of the gazelle, here a fossil gazelle from Algeria (Gazella dziria). The postural balance of gazelles is among the most harmonious of the family Bovidae, withers and hindquarters are on the same plane, the head is carried high thanks to a vertical neck (© Hadjouis and Le Bihan)*

4

Cervidae

4.1. The red deer (*Cervus elaphus*)

Order *Artiodactyla*.

Family *Cervidae* Gray, 1821.

Genus *Cervus* Linnaeus, 1758.

Species *Cervus elaphus* Linnaeus, 1758 (red or red deer).

4.1.1. *Chronological, geographical and morphological indications of species*

The red deer of Europe and Asia is part of the *Cervidae* family, known since the Tertiary period. They are animals with frontal appendages whose antlers fall off and grow back every year. The teeth are brachyodont (low and wide crown) and the metapods have their own characteristics (posterior groove on the metatarsal of the hind leg, anterior groove on the metatarsal of the front leg). The oldest *Cervidae* (*Pliocervinae*) still had lateral metapods (holometacarpalians). Other *Cervidae* that are still alive have vestigial metapods at their proximal extremities (plesiometacarpalians) (deer and buck) or at their distal extremities (telemetacarpalians) (roe, elk and reindeer). The external characteristics of the present-day red deer (coat, antler shape, tooth shape and cranial morphology) are on the whole unchanged since the Middle Pleistocene (about 500,000 years ago), except for variations in size, which in some regions may be reduced (Figure 4.1). It was present in many Middle and Upper Pleistocene sites in Europe.

Many genera and species are known in Europe since the Ruscanian (genera *Cervus*, *Croizetoceros*, *Eucladoceros*, *Arvernoceros*, *Dama*, *Capreolus*, *Alces*, *Libralces*, *Rangifer*, *Praemegaceros* and *Megaceroides*). For these last two genera, see below.

The craniofacial structure of red deer seems to be balanced between the facial and spinal skull (Figure 4.2), although a clear basic cranial flexion characterizes the *Cervidae* in general. In fact, on the deer skull in lateral view and the one raised laterally, we note the verticalization of the basi-occipital and basi-sphenoidal parts, the latter being parallel to the vault (Figures 4.3 and 4.4). This strong flexion has increased craniofacial height, which is mentioned in the closing of the upper corner of the triangle. Seen from below, the symmetry of the dento-skeletal paired parts is free of any anomaly (Figure 4.5).

Figure 4.1. *Posture of the head and quadruped body of the Barbary stag (Cervus elaphus barbarus). The posture is balanced between a front and hindquarters carried on slender limbs, the center of gravity is positioned in the middle of the body, the neck and head are above the withers (© Hadjouis and Le Bihan)*

The bone remains (antlers and bones) of the red deer are present in the Paris Basin according to frequencies related to the good or poor conservation of the sites.

Indeed, archeological interventions on the Longboyau plateau, in the Seine plain or at the Seine-Marne confluence have each time revealed the presence of sporadic or more significant remains of this deer. They are poorly represented in Paleolithic, Neolithic and protohistoric sites on the edges of the plateau (Vitry, Thiais and Rungis). On the other hand, in Holocene valley bottom sites (Ivry-Grand-Ciel, Maisons-Alfort 2) or elsewhere in the sites of the Petite Seine valley in the south of the Paris Basin, its remains were more numerous.

In the two Middle Neolithic sites, the same procurement strategy was applied in the face of a demographic increase and insufficient livestock products. Thus, in Maisons-Alfort 2, domestic and wild species did not give the impression of an imbalance between practices related to breeding and hunting activities. The Neolithics of this period and in particular in Maisons-Alfort 2 took advantage of all the animal's resources (consumption, tools made from bone and wood, tendons, etc.).

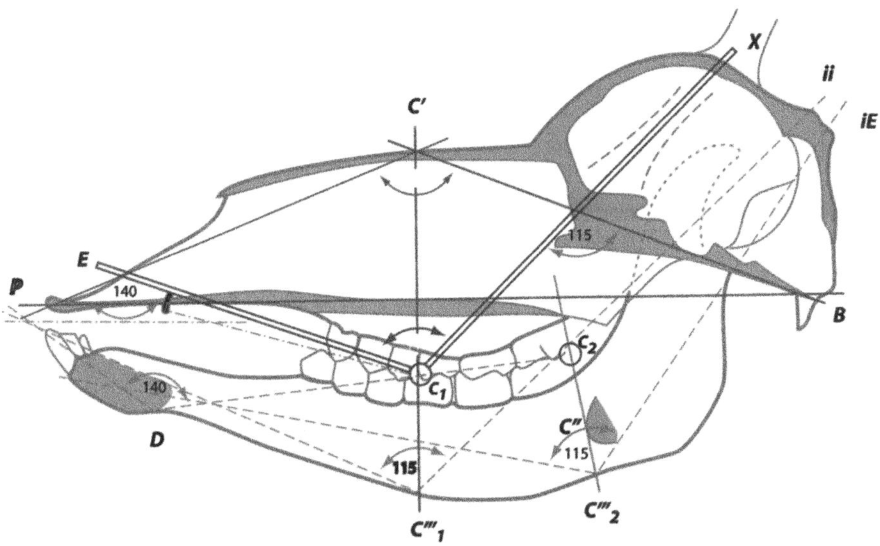

Figure 4.2. *Craniofacial and occlusal structural model of red deer. The geometrical basion-prosthion triangle shows the extension of the craniodental and craniospinal fields. The perpendicular line of the triangle passes through M1 (according to Leroi-Gourhan (1983) modified). For a color version of the figure, see www.iste.co.uk/hadjouis/skull.zip*

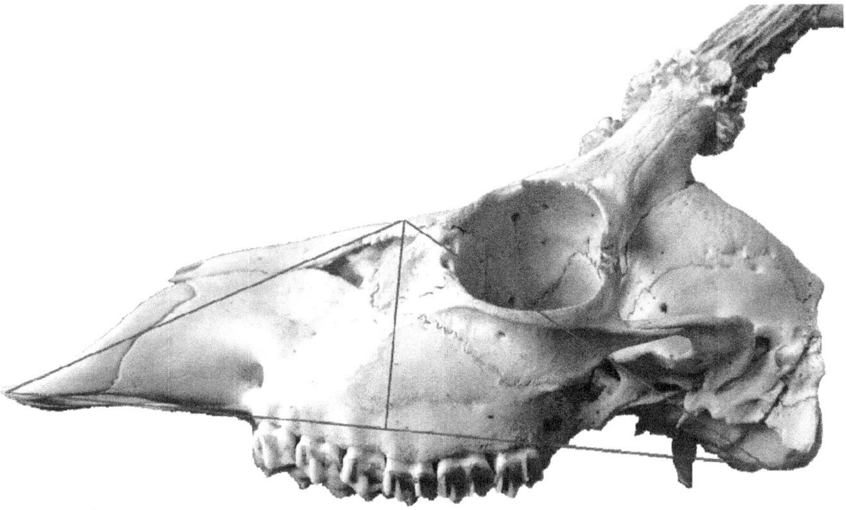

Figure 4.3. *Roe skull (Capreolus capreolus) showing a more closed triangle giving rise to a slightly stronger basic cranial flexion than red deer (© Hadjouis)*

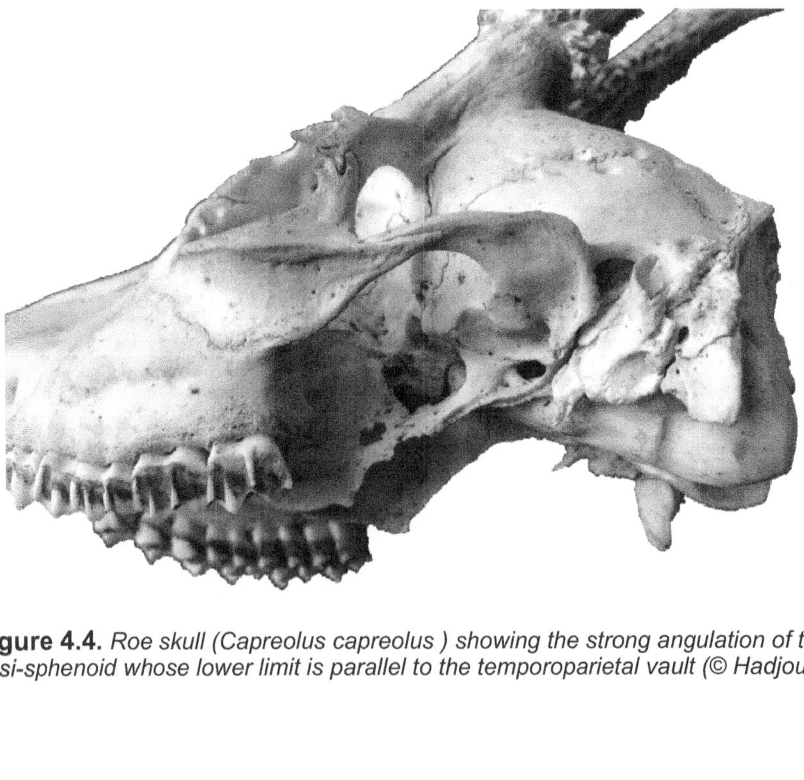

Figure 4.4. *Roe skull (Capreolus capreolus) showing the strong angulation of the basi-sphenoid whose lower limit is parallel to the temporoparietal vault (© Hadjouis)*

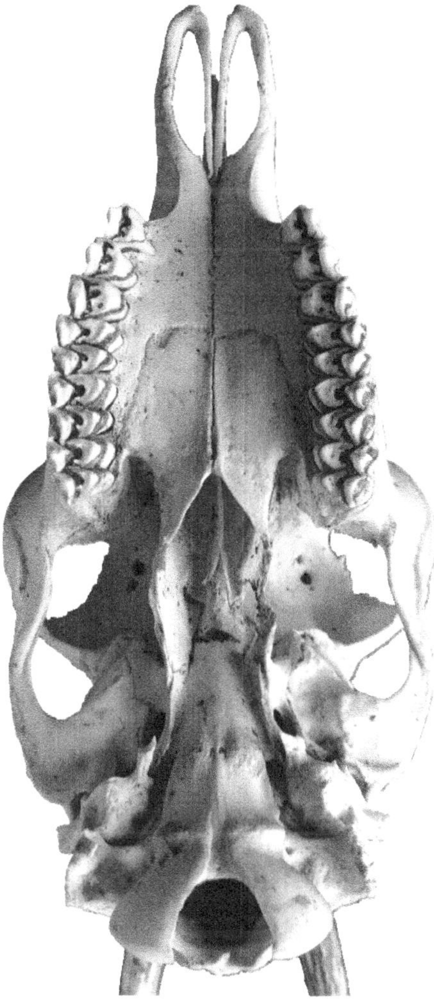

Figure 4.5. *Roe skull (Capreolus capreolus) seen from below, showing the symmetrical parallelism of the paired parts of the jugal teeth and craniofacial structures (© Hadjouis)*

Because of its status and its techno-economic value (highly symbolic species, butchery, handicraft using antlers, etc.), the red deer represents an additional value compared to the bestiary of the surrounding valleys. Because of its marked territorial behavior, selective hunting, especially that of male deer with antlers, responds to the image of the territorial stability of the populations. Indeed, scrap antlers have been sought after for their great role in agriculture and domestic life in general (axe sheaths used for deforestation, debarking or barking wood, picks, diversified tools, etc.) (Hadjouis 2008a). In addition, Neolithic occupations were marked by significant reductions in tree cover and a roughening of the sites with the installation of agricultural practices in cultivated fields, as attested by the habitats of Maisons-Alfort 2 and Ivry-Grand-Ciel. The storage of foodstuffs and an increasingly sedentary lifestyle considerably led to the increase in population.

In these sites, certain changes in pastoral activity have been identified over 1,500 years of occupation, with a significant increase in meat supply as of the Early Neolithic/Middle Neolithic transition. Hunting was essentially oriented toward the red deer, with a predominance of about 65% at the beginning of the Early Neolithic, 80% at the end of the Early Neolithic and during the first part of the Middle Neolithic and 75% at the end of the sequence.

At the Bercy site, and more precisely in the area where the southern neighborhood habitat layers of the channel portion come from, more than 100,000 bone remains have been collected. Even if much of the material was reduced to fragments, the hunting activities in relation to other domestic activities were far from negligible. Among the 13 taxa which refer to species of large and medium size frequenting various environments, it is still the red deer which predominates in the faunistic spectrum.

4.1.2. *The status of deer developing over the centuries*

In Antiquity, the social and cultural values of Gaul were modified not only in relation to the status of the city or village populations but also in relation to the food they ate, hunted or raised. Livestock production was concentrated on beef and sheep according to regional particularities. These changes, which took place with the Roman conquest, illustrate certain hunting traditions that were the opposite of what was practiced in previous periods. The red deer was no longer a prestigious animal of the forest, it was forsaken by the Romans as it was elsewhere by the Greeks. Deer hunting became despised because in the eyes of the Romans, who were used to attacking fierce animals, it was weak, timid and cowardly, fleeing from dogs (Duchet Suchaux and Pastoureau 2002).

With the Middle Ages, socio-cultural values were strongly impregnated with Christian religion and questions relating to hunting, breeding and food were subject to the statements of the sacred texts. Fortunately, the clerical discourse that opposes man created in the image of god to the inferior and submissive animal creature is not unique, and animal curiosity also evades a current of Christian authors who support the biological kinship between man and animal. In this second current, the assimilation of the Aristotelian heritage on animals (Aristotle's idea of the community of living beings) was facilitated within the Christian tradition itself.

Thus, discredited in Antiquity, the red deer regained its privileged place, some centuries later, in the hunting world. In fact, as early as the Middle Ages, the royal hunt was generalized; one of the rare hunts not to be condemned by the church, and deer game became the animal of aristocrats because it symbolized the rebirth of its constantly renewed antlers.

4.2. The Algerian thick-cheeked deer (*Megaceroides algericus*)

Order *Artiodactyla.*

Family *Cervidae* Gray, 1821.

Genus *Megaceroides* Joleaud, 1914.

Species *Megaceroides algericus* Lydekker, 1890.

For a long time in duality with the genus *Praemegaceros* Pohlig, *Megaceroides*, defined from the species type *Cervus algericus* Lydekker, 1890 (= *Megaceroides algericus*), represents today as well the African species of the Upper Pleistocene (in fact it concerns sites in Algeria and Morocco) as the European species. Although few in number, the remains of this megacerin are found especially in Algeria, notably in the travertines of Hammam El Meskhoutine near Guelma (Lydekker 1890), in Berrouaghia (Pomel 1892, 1893), Cap Carbon (Pomel 1893), Bains Romains (Arambourg 1931), Pointe Pescade and Ain-Benian (Arambourg 1932), Tamar Hat (Arambourg *et al.* 1934; Saxon 1975), Filfila (Ginsburg *et al.* 1968), Taza (Hadjouis unpublished) and the Warthogs (Hadjouis 1985a, 1990, 1994). These sites, all of which are contemporary to the Upper Pleistocene, are associated with the mustero-material and epipaleolithic industries, with the exception of the karstic fissures of the Filfila massif, which are devoid of any archeological context.

Lydekker's observations made in 1890 on the jawbone from Hammam El Meskhutin earned him the first descriptions of a megacerin taxon. Without really bringing it closer to the latter, his descriptions already alluding to the European Megaceros: "The morphology of *Cervus algericus* moves away from the characters of the molars of red deer by square and brachyodont crowns and a very prominent cingulum".

4.2.1. *Several species from Europe, the Mediterranean islands and one species from the Maghreb*

Pomel's (1892) taxonomic changes of this fossil, which he named *Cervus pachygenis* on the basis of mandibular remains, only confirmed Lydekker's validity. The detailed analysis by Joleaud (1914, 1916), on the other hand, was more in-depth, showing the great complexity of this fossil, which offered both archaic and evolved dental characters. The creation of the subgenus *Megaceroides*, characterized by the mandibular development of pachyostosis, the development of the cingulum, the accessory pillars and the external ridges of the molars, passed into the genus *Megaloceros* after multiple changes, as did *Praemegaceros*, realized by European paleontologists (Azzaroli 1953; Vislobokova 1990; Lister 1993; Van der Made 2001), which once again passed into the genus or subgenus *Megaceroides* (Ambrosetti 1967; Azzaroli 1979; Azzaroli and Mazza 1992; Abbazzi *et al.* 1999), then following the critical analysis of Hadjouis (1990) in which taxonomic changes were made (Abbazzi and Masini 1997; Abbazzi *et al.* 1999; Croitor and Bonifay 2001; Abazzi 2004; Croitor *et al.* 2006; Fernandez *et al.* 2015; Van der Made *et al.* 2017).

Thanks to the numerous dental and cranial remains of the Algerian deer found in the Aterian Warthog deposit, a complete revision could be carried out on the whole North African material and compared with the Pleistocene megacerins of Europe and Asia (Hadjouis 1990). It appears that *M. algericus* differed both from the megacerins of the Verticornis-Solihacus group of Europe and from those of the Mediterranean islands.

The cranial and dental characteristics of *M. algericus* compared with these different forms of *Megaceroides verticornis*, *Megaceroides dawkinsi*, *Megaceroides solilhacus*, *Megaceroides cazioti*, *Megaceroides cretensis*, *Megaceroides messinae*, *Megaceroides algarensis* and the Asian species *Sinomegaceros pachyosteus* show notable differences, notably a simple or molarized P4 associated with moderate pastyostosis (*M. verticornis*), or an archaic P4 associated with marked pastyostosis (*M. dawkinsi*). Similarly, the posterior tine (*M. verticornis*, *M. cazioti* and *M. algarensis*) was consistently present (Figures 4.6–4.8).

Sinomegaceros pachyosteus (Young 1932) of Choukoutien from localities 1, 9 and 13 shared synapomorphs with *M. algericus* notably in terms of the mandible (coronoid process with oval and vertical section, massive condyle, localized pachyostosis especially from the molar series and further back) and of the teeth (exaggeration of styles and pillars, very developed column, very marked cingulum and very brachyodont teeth).

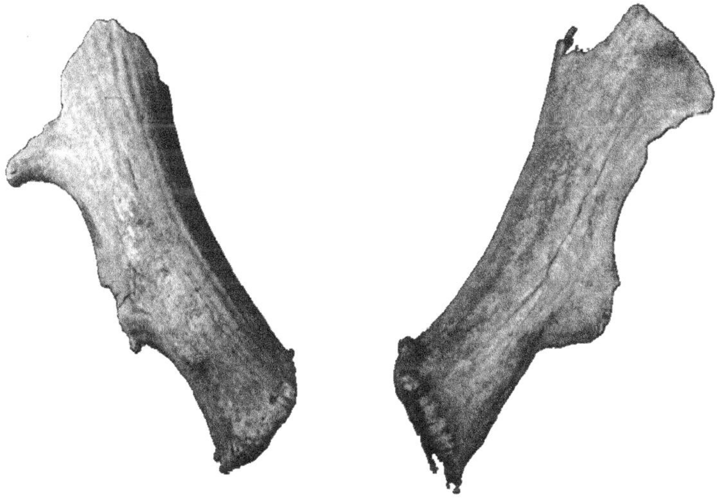

Figure 4.6. *Megaceroides algericus antler, seen from the front, from the Aterian Warthog deposit showing the position of the brown tine of antler and the (fragmentary) start of interdigital webbing (© Hadjouis)*

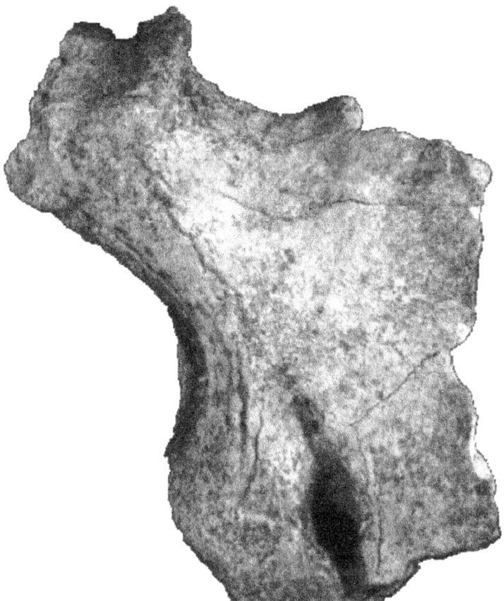

Figure 4.7. *Right frontal portion of Megaceroides algericus from the Aterian Warthog deposit showing the position of the supraorbital foramen, the eyebrow arch, a high, thick and divergent pedicle and a circular burr (© Hadjouis)*

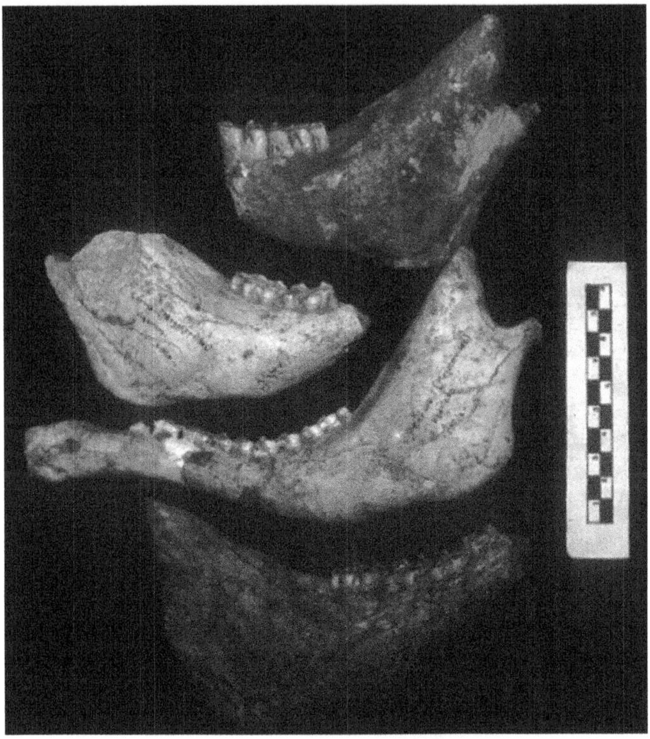

Figure 4.8. *Right and left hemi-mandibles of Megaceroides algericus from Filfila showing the strong thickening of the mandibular body due to the phenomenon of pachyostosis (© Hadjouis)*

The neodiagnosis (Hadjouis 1990) that we established following Euro-Asian and island comparisons shows a *Cervidae* species whose size is between that of the fallow deer and that of the red deer, with an entirely hyperossified skull, flattened (and not convex) forehead, short face, divergent, thick and divergent pedicle and circular burr. The lower antler (antler 2) is always located at a certain distance from the burr, with a flattened or oval section. The angle of the bez tine is formed with the pedicle close to 100°. The pedicle has an oval section before the bez tine and flattened transversely above this tine, diverging outward and backward; the trez tine being in anterior position. The second segment is twice as long as the first. There is the beginning of interdigital webbing from the trez tine. Very pachyostosed mandible, brachyodont teeth, with P4 molarized.

However, Kroitor (2016) in a recent study took up the problematic issue of the classification of *M. algericus* and proposed an amended diagnosis, in which he reviewed some criteria related to antlers and teeth, classifying it in the genus *Megaceroides*.

4.2.2. *Size of* Megaceroides algericus

Long considered as a dwarf species whose size is close to that of the island species of *Megaceroides* (*M. cazioti*, *M. cretensis*, *M. messinae* and *M. algarensis*), the Algerian species has isolated itself in the coastal territories of Algeria and Morocco, leaving its ancestral Euro-Asian origin. Its speciation as a result of genetic drift could explain the changes that the species has undergone in the Maghreb (Figure 4.9). However, the transformation, particularly of the ramifications of the antlers and the reduction in size, is not exclusive to this megacerin, since another *Cervidae* species has found refuge in the same African territories, following the example of *Cervus elaphus barbarus*. This subspecies of red deer is characterized by a smaller stature, a reduction in the size of its antlers and a reduction in the number of antlers.

Figure 4.9. *Reconstitution of the antlers of Megaceroides algericus on a body whose size has been reduced after a genetic drift from Europe (© Hadjouis and Le Bihan)*

5

Suidae

5.1. The wild boar (*Sus scrofa*)

Family *Suidae* Gray, 1821.

Sub-Family *Suinae* Gray, 1821.

Genus *Sus* Linnaeus, 1758.

Species *Sus scrofa* Linnaeus, 1758 (wild boar).

5.1.1. *Chronological, geographical and morphological indications of species*

The fossil form is robust compared to the current species. The canines of the male are powerful and the jugal teeth are large.

Suinae are characterized by the presence of a mobile snout at the end of the muzzle, supported by two nostril bones that attach to the nostrils, to the premaxillaries and sometimes to the mesethmoid (Lecointre and Le Guyader 2001). The jugal teeth are brachyo-bunodontal.

The dental formula is: 3/3 I, 1/1 C, 4/4 Pm, 3/3 M.

The upper molars have no hypocone and paraconule. They are four-folded main tubercles and several accessory tubers. The lower molars are free of paraconids and also have four tubercles and main and accessory tubers. Premolars do not show any form of molarization (Guérin 1996c).

Suidae appeared in the Upper Miocene with forms such as *Listriodon* with archaic molars, bunodonts for some (*Bunolistriodon*) and lophodonts for others (*Listriodon*). The modern *Suidae* first appeared in the Pliocene with *Potamochoerus*, close to the present African species, while the genus *Sus* attested in the Plio-Pleistocene (Guérin 1996c). Fossil species of this genus are known as *Sus arvernensis* Croizet and Jobert, whose last molars ended in a single tubercule, *Sus strozzii* Meneghini, of great size, *Sus scrofa* Linnaeus, the wild boar of Eurasia and *Sus nanus* Van der Made, an island species of Sardinia, were attested to since the Upper Ruscinian for the first of them, up to the Middle Pleistocene. Sites are numerous such as Vialette, Perrier-les-Etouaires, Saint-Vallier, Senèze, in the French Massif Central, the Arago Cave, Lunel-Viel, Vallonnet (France), Val di Pugna, Pieve Fosciani in Tuscany and in Piedmont (Triversa), Forest-Bed (England), in Mosbach, Mauer, Süssenborn, Voigtstedt and Undermassfeld (Germany).

Neolithic sites south of the Paris Basin have presented different types of fishing hooks fashioned from boar teeth.

For the pig (*Sus domesticus*), among its cranial particularities are: a temporal bone already unified at birth and rapidly fusing with neighboring bones; a remarkably powerful zygomatic process, bent at right angles; very thick parietal bone, fused to its homolog from the age of six months and strongly tilted backward against the occipital; very extended maxilla; massive and powerful mandible, fused very early to its homolog (Barone 1976).

Like cattle, pigs are a species domesticated in France from an ancestral wild boar strain from Eurasia. This phenomenon has moreover led to certain difficulties in distinguishing between the skeletal remains of boar and those of pig, particularly in the Early Neolithic, due to the fact that the size and morphology of the two species were still identical at the time of the first generations. In the Paris Basin, pig's food during the Neolithic often served as a supplement after small ruminants, sometimes in a secondary situation after cattle. Classification depends on the importance of sites in rural, peri-urban or urban areas, especially from the Tène and Gallo-Roman periods. In Val de Marne, all the sites that have provided more or less consistent wildlife numbers place pigs in third place after sheep and/or goats in the early Neolithic periods, sometimes in second place in the Final Bronze Age. The preferential slaughter of males was generally before 2 years of age and a little more for females.

Elsewhere in some sites in Seine-et-Marne, the management of pigs in the Final Bronze Age was much more important and took priority over cattle and goats and pigs. Slaughter was quite variable and could vary from 6 to 12 months, even to 60 months in sites such as Boulancourt.

The main species domesticated at the beginning of the Holocene that have no wild ancestors in Western Europe come exclusively from the Near East. This is the case for goats and sheep. The latest studies on the birth of animal species breeding in domestication sites no longer point to a strong technical innovation by the societies concerned, but rather to one of the manifestations of their profound cultural change. To this end, the movements of communities with their livestock were not based on massive migratory transfers. Archeozoological data show a gradual spread with sometimes an acculturation of neighboring societies but not always. As these transfers were a choice of the Neolithic communities, certain regions in Scandinavia and the Baltic countries retained the mode of predation and that goats and sheep only made their entry late between the 1st and 2nd millennium BCE.

Concerning the species raised with a local wild ancestor in certain geographical territories, as is the case for the canine racial diversity of the dog (from the wolf), the cattle varieties (from aurochs) or the pig varieties (from wild boar), several hypotheses are indeed in favor of on-site domestication.

5.1.2. *The status of the boar over the centuries*

The boar occupied a privileged place among the Romans. Unlike the red deer, the boar was for the Romans noble game whose strength and courage led to admiration, and this symbolic tradition remained until the Middle Ages, above all in the Germanic countries. Prolific in archeological sites is the hunted part of the boar.

In France, this tradition was lost from the 13th century onward and at the end of the Middle Ages; boar hunting was rather a simple hunting practice to eliminate prolific individuals that ravaged the vineyards (Duchet Suchaux and Pastoureau 2002).

5.1.3. *Postural balance of the boar*

The wild boar has a different silhouette from the warthog, but similar to the bushpig and forest hog. While the warthog has a structural antero-posterior axis with the front and hindquarters forming like the two columns of a suspension bridge, the cepahalo-caudal axis of the wild boar is in a straight line with no central concavity corresponding to the thoracolumbar spine, but with an elevation of the part corresponding to the thoracic segment (Figure 5.1). The cervical spine is lower than the thoracic spine and in a straight line, maintaining a low head position. The head is relatively long and pointed and is equipped with short tusks. The withers and the hindquarters are carried by powerful and robust limbs.

Figure 5.1. *Posture of the head and quadruped body of the wild boar (Sus scrofa scrofa). In spite of a distinctive silhouette, the body is balanced between front and hindquarters, the neck and the head are positioned lower than the hindquarters (© Hadjouis and Le Bihan)*

The boar's craniofacial structure shows a greater anterior-posterior extension noted in the opening of the corner of the upper triangle. The length of the skull is reserved for the greater part of the face and there is little flexion making the basi-occipital and basi-sphenoidal floor in an axis close to the palatal bone (Figure 5.2). This configuration has given rise to a *foramen magnum* positioned in the same horizontal direction (Figure 5.3). In the lower view, the symmetry of the even parts is evident. All the dental series of the half jaw (incisors, canines [defenses], premolars and molars) face the antagonistic series perfectly equilibrated and no dento-skeletal anomalies seem to be apparent (Figure 5.4).

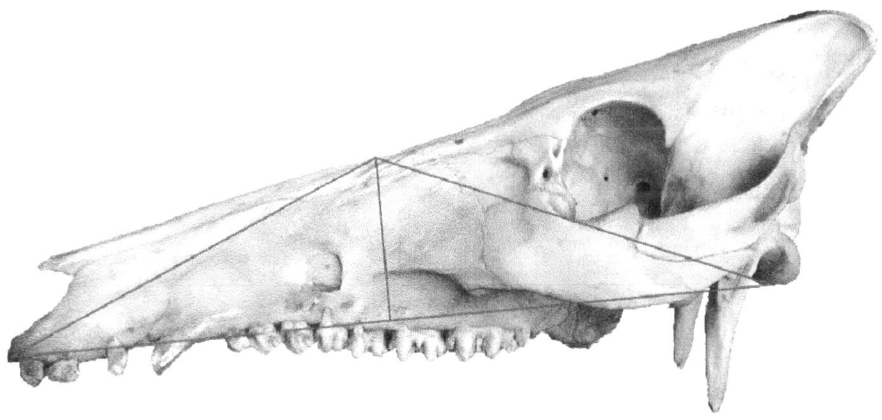

Figure 5.2. *Boar skull (Sus scrofa scrofa) in left lateral view. The geometric basion-prosthion triangle shows the extension of the craniodental and craniospinal fields. The perpendicular line of the triangle passes between M1 and M2 (© Hadjouis)*

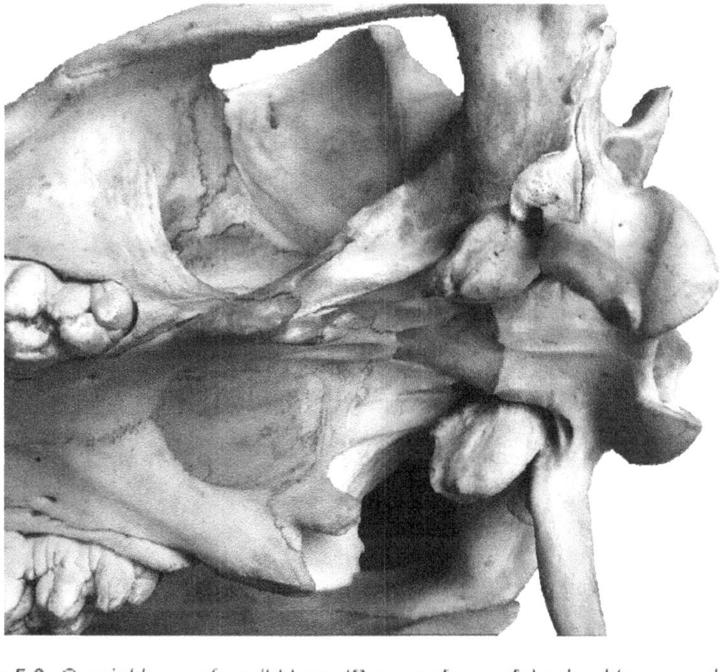

Figure 5.3. *Cranial base of a wild boar (Sus scrofa scrofa) raised to appreciate the weak flexion making the basi-occipital and basi-sphenoidal floor be in an axis close to the palatal bone (© Hadjouis)*

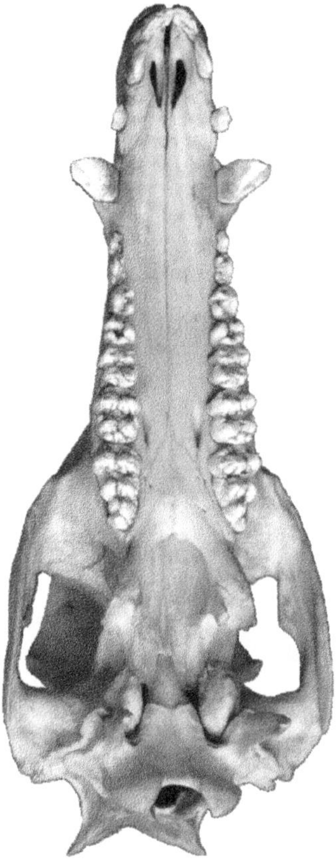

Figure 5.4. *Dental and craniofacial balance in a boar, seen from below. The incisal, canine, premolar and molar rows are aligned mesio-distally, and the even parts of the zygomatic arches, the articular tubercles of the temporal bone, the jugal processes, etc., are balanced and free of asymmetries (© Hadjouis)*

5.2. The warthog (*Phacochoerus aethiopicus* or *africanus*)

Family *Suidae* Gray, 1821.

Sub-Family *Phacochoerinae* Van Hoepen, 1932.

Genus *Phacochoerus* Cuvier, 1817.

Species *Phacochoerus africanus* Gmelin, 1788 (common warthog).

Species *Phacochoerus aethiopicus* Pallas, 1767 (desert warthog).

5.2.1. *Chronological, geographical and morphological indications of species*

Two Sub-Families of *Suidae* are represented in the African continent: the *Suidae*, of Eurasian origin, with the genera *Sus* and *Potamochoerus* and the Phacochoerines, exclusively African, represented by *Phacochoerus aethiopicus* and *Hylochoerus*. Concerning the Suina, and in particular the genus *Sus*, it only appeared in North Africa during the Upper Pleistocene.

The profusion of fossil genera and species of *Phacochoerinae* (*Nyanzachoerus*, *Notochoerus*, *Kolpochoerus* and *Metridiochoerus*) has made their classification difficult because of their low diversity, especially since determinations are made on even poorer materials in the Quaternary deposits of North Africa.

The separation of the two Sub-Families is based on the type of dentition, brachyodonts with persistent premolars in adults and molars free from cementum in *Suinae*, hypsodonts or semi-hypsodonts, partially deciduous premolars in adults, reduced incisors and molars particularly free from cementum in *Phacochoerinae* (Arambourg 1947).

According to zoologists, *P. aethiopicus*, a species recently extinct in Africa, south of the Sahara, is the most common and the most cited in inventories, while in the faunal lists of Quaternary deposits in the Maghreb, we find sometimes the form *africanus*, sometimes the form *aethiopicus*.

Lönneberg (1908), Shaw (1939) and Aramburg (1947) insisted on the distinctive character of the teeth of these two species. The desert warthog is considered more progressive, with in particular a disappearance of the incisors but especially a greater hypsodontia of the molars whose pulp cavities remain open. As for the *africanus* form, it kept its lower incisors and the pulp cavities close before their total eruption (Arambourg 1947).

Work on African *Suidae* (White and Harris 1977; Harris and White 1979) devotes only a small part to the genus *Phacochoerus* and only includes the Ethiopian species. These authors consider, on the other hand, that *Phacochoerus antiquus* Broom (1947) from South Africa is the direct ancestor of the modern forms.

The warthog's craniofacial structure is similar to that of the wild boar, although the latter has a greater antero-posterior length (Figure 5.5). The present species is characterized by its imposing upper tusks (Figures 5.6 and 5.7), whose excessive growth sometimes results in oral pathologies.

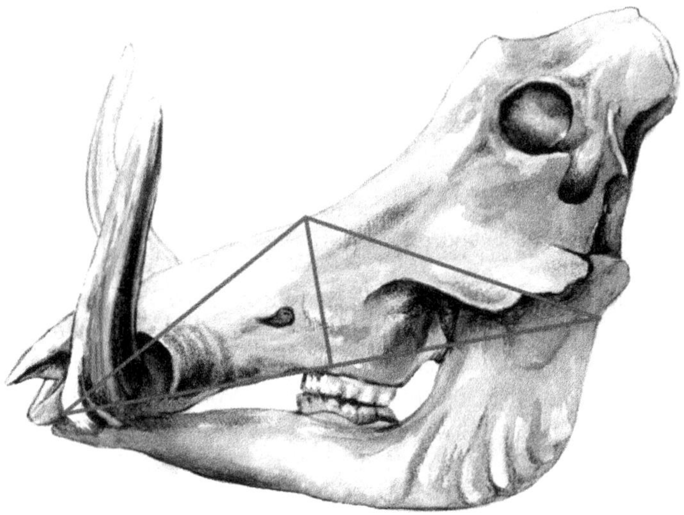

Figure 5.5. *Present-day warthog skull (Phacochoerus aethiopicus) in left lateral view. The geometric basion-prosthion triangle shows the extension of the craniodental and craniospinal fields. The perpendicular line of the triangle passes through the single M3 of the half jaw (© Hadjouis and Le Bihan)*

Figure 5.6. *Present-day warthog skull in left lateral view showing the occlusion in the presence of the only third molars after pushing the anterior jugal teeth (courtesy of J.-D. Vigne, Comparative Anatomy of the MNHN, © Hadjouis)*

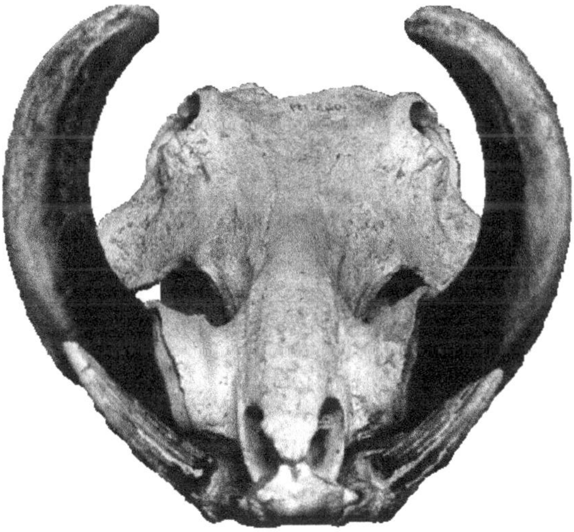

Figure 5.7. *Present-day warthog skull in facial view showing the excessive development of the upper tusks (courtesy of J.-D. Vigne, Comparative Anatomy of the MNHN, © Hadjouis)*

The Aterian Warthogs deposit located on the heights of Algiers (Algeria) has brought to light more than a hundred skeletons of all ages and remains. To this day the Quaternary site has yielded the largest number of fossil warthogs in the African continent (Hadjouis 1994, 2003a, 2003d; Hadjouis and Roubet 2020). This number remains much lower than the original site, which was considerably reduced due to the conditions of salvage excavations. In addition to tusks (Figure 5.8) in the set of this warthog's jugal teeth are a large number of molars, notably M3. This distinction between the two species does not appear and the closure or not of the pulp cavities is only related to age variation.

In the dental material of the Algerian deposit, the M3 of the very old individuals closed their pulp cavities while reducing their height, and it is only on this characteristic that this closure was applied. In this species, apart from the very advanced age, the jugal teeth in particular the M3 remained in their great majority open. However, and this is a general rule that applies to this type of eruption, the distal roots closed earlier than the mesial ones due to a greater reduction in height in its posterior part as wear progressed. By this characteristic, the Algerian species and probably by expansion the North African species could only be compared to the current species *aethiopicus* which presents with one notable difference, except for a larger size. As for the highly developed upper canine teeth, they were similar to those of the present-day desert warthog.

Figure 5.8. *Tusks (canines) of fossil warthogs (Phacochoerus aethiopicus) found in the Aterian site of the same name (© Hadjouis)*

5.2.2. *A particular tooth eruption*

The warthog adopts a peculiar tooth eruption that is somewhat similar to that of the Proboscidians, especially in the eruptive mode of the molars. While premolars experience a classic replacement mode of eruption in a vertical eruption, this is not the case for molars. The molars adopt a horizontal eruption from back to front, and it is precisely the third molar with a progressive mesio-distal diameter of the occlusal surface that will take up all the space (Figures 5.9, 5.10 and 5.11). The appearance of this back molar, its hypsodont character and its very great length, which will occupy all the alveolar sites of the jugal series, after having displaced the premolars and molars (M1, M2), recall by these characteristics the dental replacement of *Elephantidae* (Hadjouis 2003c, 2003d). This characteristic was already present in *Metridiochoerus compactus* of the Middle Pleistocene.

Tooth eruption is completed at about 4–5 years of age and is carried out as follows:

– from 0 to 1 year: in the maxilla, eruption of the canine, eruption of P1 (it can also be absent) of D2, D3 and D4 and of M1 (around 7 months), appearance of M2. In the mandible, P1 and P2 are already erupting after the canine, loss of P3, appearance of P4 and M2, eruption of M1;

– from 1 to 2 years: in the maxilla, loss of P1 and P2 and eruption of P3, P4, M1 and M2, appearance of M3. In the mandible, eruption of M2 and appearance of M3;

– from 2 to 3 years: in the maxilla, remains of M1, eruption of M3. In the mandible, remains of M1, eruption of M3;

– from 3 to 4 years: in the maxilla, loss of M1, very worn M2. In the mandible, loss of M1, remains of M2. The growth of the upper canine exceeds this age and can sometimes extend to around 10 years.

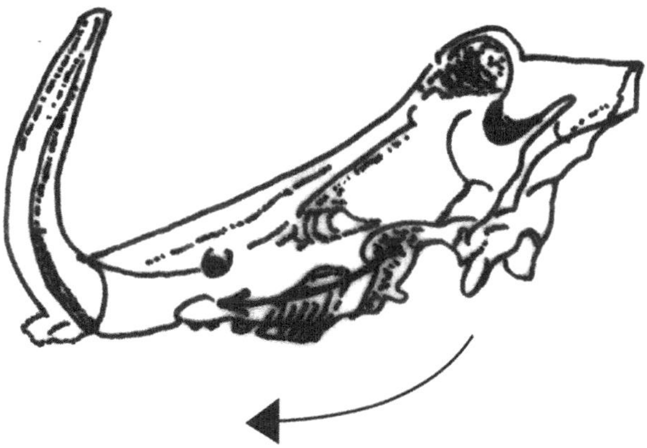

Figure 5.9. *Replacement of jugal teeth in warthogs. Premolars and first molars are expelled from back to front, keeping only the M3 in each half jaw (© Hadjouis)*

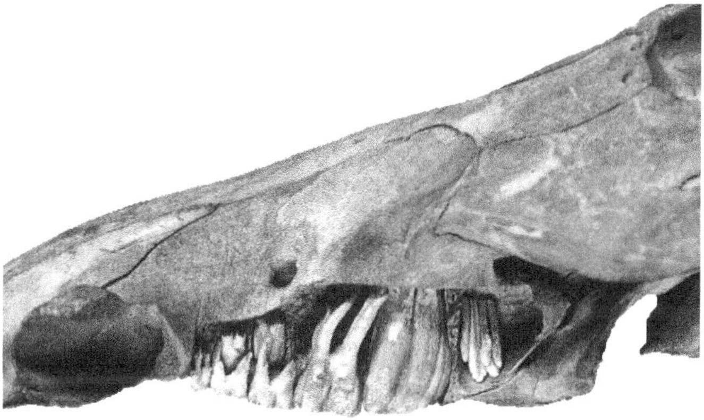

Figure 5.10. *Skull of a young warthog, 7 months old, in left lateral view showing the eruption of the upper jugal teeth (courtesy of J.-D. Vigne, Comparative Anatomy of the MNHN, © Hadjouis)*

COMMENT ON FIGURE 5.10. – *While the replacement of premolars is the same as that of all Mammals (vertical eruption), the molars undergo a horizontal push from back to front (here M1 and M2 are one behind the other), the latter will take up all the space after driving out M1 and M2.*

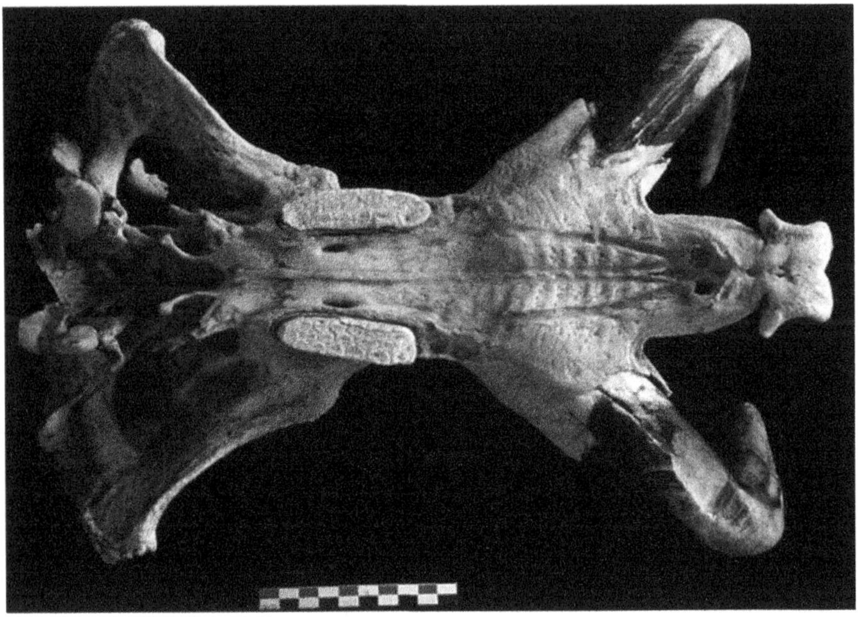

Figure 5.11. *Present-day warthog skull from below showing the unique presence of the third molars after pushing the anterior jugal teeth out. Note the clearly visible wear of the upper tusks made by the lower antagonists (courtesy of J.-D. Vigne, Anatomie comparée du MNHN, © Hadjouis)*

5.2.3. *Postural balance of the warthog*

Although it is from the same Family *Suidae*, the warthog does not resemble its cousins, the bushpig, the forest hog or the wild boar. Its balanced silhouette and high head carriage give it the appearance of a stocky equine with a massive skeleton whose limbs show the robustness and massiveness of fossil bones. Indeed, it is especially by the position of straightened hindquarters and an elongated body that this comparison is possible. The head, although very large, remains proportional to the dimensions of the body whose characteristic is an animal with hooves higher than the forest hog, but of a size close to that of a wild boar (about 75 cm at the withers). The head is positioned high thanks to a verticalization of the neck spine from the withers. These two essential characteristics (postural balance between the front and hindquarters) as well as a raised head carriage make the warthog a special species in the *Suidae* family.

5.2.4. *Pathologies in warthogs*

Two forms of pathologies are usually found in this animal. The first one is directly related to the excessive development of its tusks. Less spectacular than the development of the canines in the Babirusa, whose upper teeth pierce the upper lip and push upward in the shape of a crescent, they nevertheless wreak havoc in the warthog's mouth. Indeed, the upper canines grow by curving upward reaching an average of 30 cm, the shorter lower ones wear out against the upper ones giving the lower tusk a sharp angle. But their coronal development is without pathological effect, it is rather the part embedded in the alveolus (in Suina, crown and root of the canine are not distinguished) that makes an excessive distal development to the point of forming another alveolus or widening the M3 alveolus to be able to escape (Figure 5.12). This continuous growth of the defenses is considered to be pathological development.

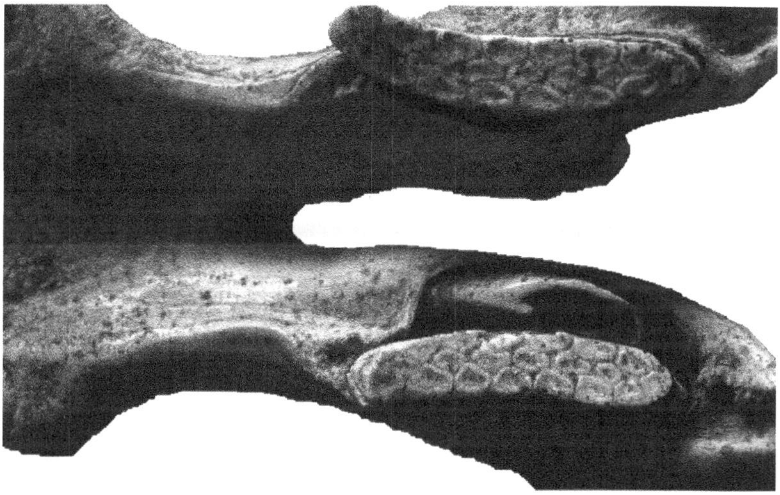

Figure 5.12. *Present-day warthog mandible seen from above showing the third molars in their alveoli (courtesy of J.-D. Vigne, Comparative Anatomy of the MNHN, © Hadjouis)*

COMMENT ON FIGURE 5.12. – *The left alveolus is more widened in its lingual part by the excessive radicular thrust of the defense. This pathological development of the tooth could hurt the anima's palate if it was not for the stopping of the upper molecular as shown by the rather pronounced wear of the canine.*

The second type of pathology encountered in warthogs is infectious and arthropathic. The species develops several forms of osteitis similar to osteoperiostitis and/or osteomyelitis on the limb bones (Figure 5.13).

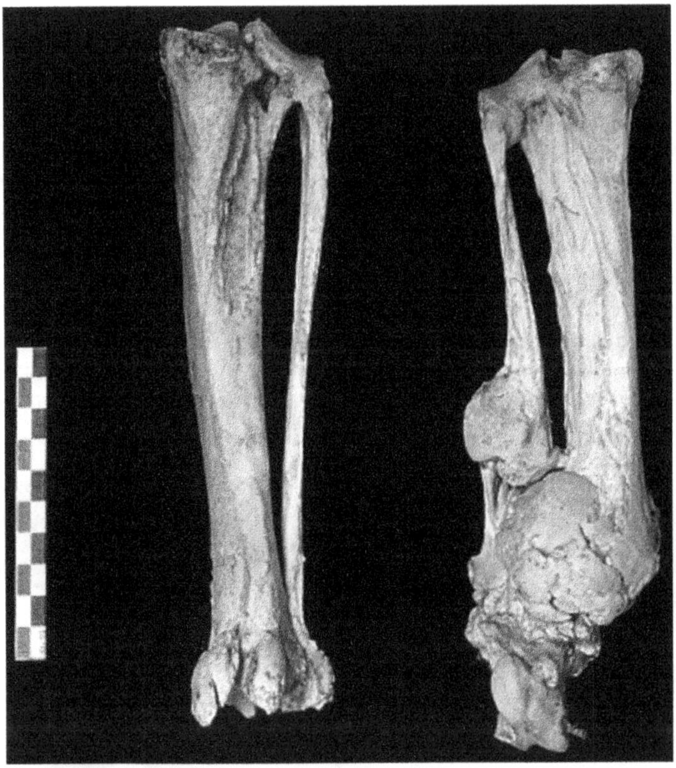

Figure 5.13. *Left leg skeleton of a present-day warthog affected by a form of osteitis (right photo) in comparison with the healthy leg on the right (courtesy of J.-D. Vigne, Comparative Anatomy of the MNHN, © Hadjouis)*

5.2.5. *A catastrophic mortality curve*

In the Aterian Warthog site highlighted above, mortality in this disease affected the entire population without exception, from newborns to very old individuals. The exception was two bones (humerus and femurs) that showed traces of tools at the distal diaphysis, the other bones, which were mostly complete, did not show the characteristics of a hunting activity and the mortality curve was not similar to a hunting curve, but to that of a living natural population.

Here the age profile was that of a catastrophic curve. The high mortality among the young and young adults between 0 and 2 years of age and among older individuals, whose age ranged between 10 and 15 years, was reminiscent in its profile of the natural mortality of warthogs in Rwandan parks and the Klasies deposit in South Africa (Klein 1979, 1982). In the latter site, warthogs were killed by carnivorous predators such as lions. In the Algerian deposit, although lions, panthers, jackals and especially hyenas were present, their dental traces were less represented on the bones of warthogs, whereas they were widely represented on the bony parts of other large mammals (Hadjouis 1985a, 1994, 2003d; Hadjouis and Roubet 2020). As for the presence on two bones of traces of anthropogenic activity (cutting and fragmentation), it suggests several possibilities. In the first, warthogs were hunted by large carnivorous predators such as the lion or panther. The Aterians, behaving like scavengers and taking advantage of these animals shredded by predators, recovered some carcasses that they skinned, thus selecting skulls for brains and limb bones for the rest of the flesh and marrow. This hypothesis shows the passage of the Aterians as scavengers after the large active predators and before the scavenging animals. The second possibility is based on the very nature of the food chain, following a sequence of stages observed in the Serengeti and Ngorongoro parks. Large and medium-sized mammals are first hunted by lions and panthers (typical predators), then hyenas (typical scavengers) finish their meal after the carcasses have been abandoned by active predators. It is also possible that the Aterian men only intervened at the third stage of the food chain, after scavengers. In this case, the presence of both cutting marks and teeth marks and chewing by predators on the same bones would be consistent with these hypotheses (Hadjouis and Roubet 2020).

Another possibility is envisaged, which opposes those seen previously and concerns the catastrophic hypothesis, also shown by the age curves (Hadjouis 1994, 2003). It could be envisaged that the high mortality of warthogs is due to death by drowning, disease or malnutrition. Bones without pathological lesions (absence of trauma, degeneration, inflammatory or infectious diseases) are hardly attributable to this cause. However, the ingestion of toxic products or malnutrition could reflect the causal element of this massacre. But then why does this hypothesis not apply to all Mammals? What then can we say of an epidemic affecting only Suina animals, as in the case of an infection similar to swine fever, knowing that the vector of such an infection does not leave visible traces on the bones.

With the main exception of microbial transmissions such as treponematosis (*Treponema pallidum*) or plague (*Yersinia pestis*), other microbial or viral diseases, transmitted to fossil species organisms, leaving no bone-marks are rarely described. Only domestic or wild species that are still alive are studied within the framework of veterinary research programs or in large African savannah parks.

6

Carnivores

6.1. The lion (*Panthera leo*)

Order *Carnivora*.

Family *Felidae* Gray, 1821.

Genus *Panthera* Oken, 1816.

Species *Panthera schaubi* Viret, 1954.

Species *Panthera gombaszoegensis* Kretzoi, 1938.

Species *Panthera (Leo) mosbachensis* Dietrich, 1962.

Species *Panthera (Leo) spelaea* Goldfuss, 1810.

Species *Panthera pardus* Linnaeus, 1758.

Species *Panthera (Leo) leo* Linnaeus, 1758 (lion).

6.1.1. *Chronological, geographical and morphological indications of the species*

The species of the *Sub-Family Felinae* are mostly homogeneous: globular head, powerful jaws, mandible without lateral movements, reduced dentition, powerful canines and secodont jugal teeth (Argant 1996b).

The genus *Panthera* gathers large and massive felines, represented in Europe by five species.

Panthera schaubi, whose size seemed to be the smallest, was close to the forms of panther. It was found in Middle Villafranchian sites such as Saint-Vallier and the Puebla de Valverde. *Panthera gombaszoegensis* was a large cat with strong canines. Its geographical distribution swept across Western and Central Europe and extended to the limits of Tajikistan, where it was present in Upper Villafranchian sites (Upper Val d'Arno, Olivola, Tegelen, Erpfinger, etc.), with persistence into the Middle Pleistocene (L'Escale, Mosbach, Uppony and Château) and possibly into the late Middle Pleistocene, at least in Central Europe (Argant 1996b). Some authors (Kurtén and Anderson 1980) believe that this feline was closer to jaguars than to lions, and that the present jaguar may represent a Middle Pleistocene form of descent. *Panthera mosbachensis*, an ancestral form of the cave lion, considered to be the most powerful of the felines, had a very large size and massive skeleton. One of the dental characteristics was the shape of M1, wide and massive, with a protoconid clearly exceeding the paraconid in height and the presence of a talonid at the back of the tooth. The remains of the cave lion's ancestor were found during the early Middle Pleistocene in Western and Central Europe at the sites of Mauer and Mosbach (Germany), Cromer and Forest-Bed (England), Atapuerca (Spain), Château, la Caune de l'Arago (France), Sprimont (Belgium), Petralona (Greece) and Verteszöllös (Argant 1996b).

Panthera spelaea is considered to be the evolutionary species of *P. mosbachensis* with characteristics that differ from the current subspecies *Panthera (Leo) leo*: a large size, large extension of the infratemporal fenestra and a reduced face size. The stage of the species was acquired with the Middle Pleistocene and lasted until the end of the Tardiglacial. It was found in the Lunel-Viel site (France) and in some paleontological sites.

The genus *Panthera* seems to have appeared in the African continent during the Lower Pliocene, accompanied by other carnivore taxa, particularly in East Africa (Laetoli). As for the species *P. leo*, of Indo-Ethiopian affinity, it seems to have appeared in the Upper Pliocene between 2.8 and 2.4 million years ago, notably in Sterkfontein 4 in South Africa.

Probably present in Aïn Hanech (near Sétif, Algeria) by a single calcaneum, the oldest Pleistocene record from Algeria, this species was, on the other hand, present in the coastal sites of the Upper Pleistocene in the Algiers region (Aïn Benian, Warthog) (Hadjouis 2003d).

6.1.2. *Occlusal posture and the lion's balance on the ground*

Like the majority of its fellow carnivores of the Order of Carnivores, but especially of the family *Felidae*, the lion is highly specialized, characterized by a rounded head, with a reduced number of jugal teeth, developed canines and sharp molars. The body of this large, very robust felid is perfectly balanced, with a horizontal cephalo-caudal axis along its entire length, which can reach 2.5 m without the tail. The cervical spine does not exceed the withers so that the head is at the same level. The limbs are large, giving the animal a high sitting position that allows it to use high speeds.

The craniofacial structural example is taken here on the basis of a wildcat's skull (*Felis sylvestris*) which demonstrates the characteristics of the *Felidae*. The skull is generally low, with a large width characterized by the development of the zygomatic arches and a reduced face (Figure 6.1). The basi-occipital and basi-sphenoid floor, which can be observed by the lateral lifting of the skull, is almost flattened, almost parallel to the palatal axis. This basic cranial configuration places the *foramen magnum* in a horizontal axis (Figure 6.2).

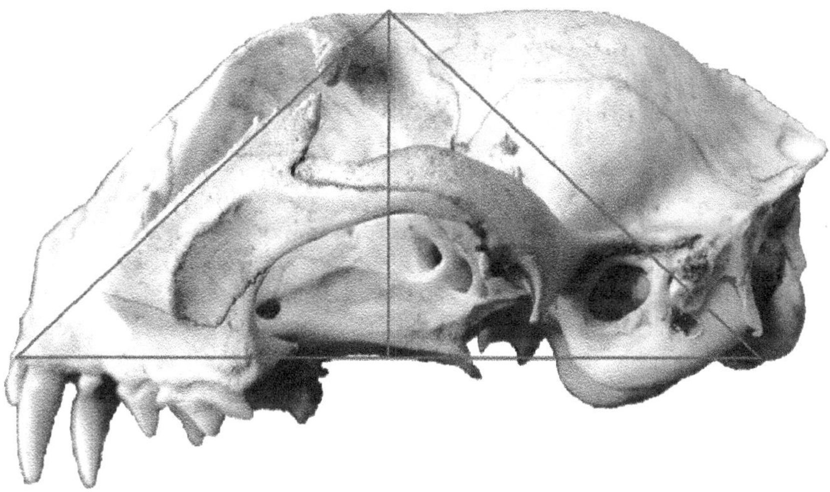

Figure 6.1. *Wildcat skull (Felis sylvestris) in left lateral view (© Hadjouis)*

COMMENT ON FIGURE 6.1. – *The geometrical basion-prosthion triangle reflects the extension of the craniodental and craniospinal fields. The perpendicular line of the triangle passes behind the temporal line and through the middle of the zygomatic arch. The lowering of the arch and the widening of the skull could be related to a zero craniofacial flexion. The foramen magnum is a continuation of a flattened basi-occipital and basi-sphenoid.*

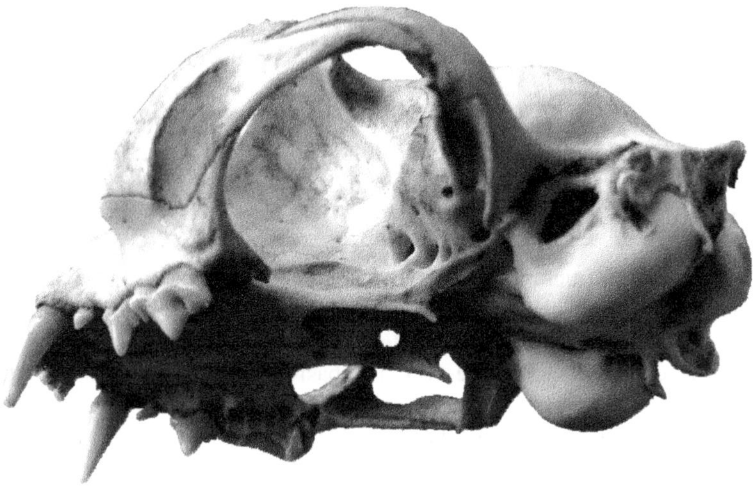

Figure 6.2. *Wildcat skull (Felis sylvestris) raising the base of the skull to appreciate a basi-occipital and basi-sphenoid floor without flexion (© Hadjouis)*

Like the boar, the status of the lion occupied a privileged place in Antiquity. The king of the animals was introduced by the Romans for circus games coming essentially from North Africa. The practice of training lions in circuses and the ferocious exhibition of these wild animals from Antiquity to modern times have taken on such a dimension that emblematic media and other codes have permeated. First names constructed from the root *Leo* (Leo, Leonardo and Leopold) or surnames (Lionnard and Leonelli) or to illustrious characters (Robert the Lion of Flanders and Richard the Lionheart) have flourished. Whereas from the 12th century, all of Europe had the lion as the main figure in medieval coats of arms (Duchet Suchaux and Pastoureau 2002).

6.2. The panther or leopard (*Panthera pardus*)

Family *Felidae* Gray, 1821.

Genus *Panthera* Oken, 1816.

Species *Panthera pardus* Linnaeus, 1758 (panther or leopard).

6.2.1. *Chronological, geographical and morphological indications of species*

Of Indo-Ethiopian affinity, the panther made its appearance in the Upper Pliocene between 2.8 and 2.4 million years ago. It was found for the first time in North Africa in the Moroccan site of Ahl al Oughlam along with other new carnivore species.

Absent from the Algerian Lower Pleistocene sites, it reappeared during the Middle Pleistocene but its remains were found especially in the Upper Pleistocene, notably in the caves around Algiers (Anglade and Sintes quarries in Ain Benian), in the Warthogs site where carnivores were more abundant, probably at the Roman Baths and in the caves of the Bedjaïa region (Tamar Hat) (Hadjouis 2003d; Olive 2005).

The current African species is approaching European forms despite the separation of these felids into two groups (Ballesio 1975). Phylogenetic analysis of current felids tends to place *Neofelis* in a sister group to *Panthera* and *Panthera uncia*, which represent the first divergence within the genus (Salles 1992). Some dental characteristics, such as the weak development of the parastyle of the third upper premolar, allow us to place *Panthera pardus*, *P. leo* and *Panthera onca* in the same group (Olive 2005). The first two species are also close according to molecular data (Johnson *et al.* 1996), while biometric data differentiate *P. pardus* from the other species, notably by the relative development of the lower third premolar (Olive 2005). In Europe, this medium-sized felid was present at about 100 sites in the Middle and Upper Pleistocene.

6.2.2. *Occlusal posture and postural balance of the panther on the ground*

This medium-sized and robust feline has short legs, suitable for jumping and climbing. Like the lion, it has highly specialized characteristics, including a rounded head and a reduced number of jugal teeth. The cephalo-caudal axis is in a postural balance allowing the hindquarters to be the same level of the frontquarters. The more vertical cervical spine positions the head above the withers (Figure 6.3).

As mentioned in the previous pages concerning the example of the cat's craniofacial, the leopard's structure follows exactly the same model (low skull, reduced face, flattened basi-occipital and basi-sphenoid floor, horizontal *foramen magnum*) (Figure 6.4).

Figure 6.3. *Posture of the head and withers of the leopard's quadruped body (Panthera pardus). Like the majority of Felidae, the hindquarters and withers are in a perfectly balanced axis, with the head in extension to the horizontal line of the back. The highly developed canines are located in arches reduced in length in response to the reduction of the jugal teeth (© Hadjouis and Le Bihan)*

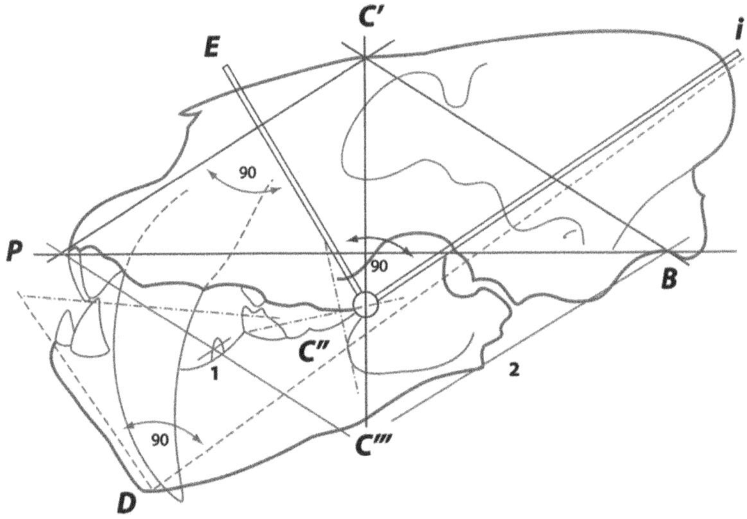

Figure 6.4. *Craniofacial and occlusal structural model of Felidae. The geometrical basion-prosthion triangle reflects the extension of the craniodental and craniospinal fields. The perpendicular line of the triangle passes behind the molars (according to Leroi-Gourhan (1983) modified). For a color version of the figure, see www.iste.co.uk/hadjouis/skull.zip*

6.3. The spotted hyena (*Crocuta crocuta*): chronological, geographical and morphological indications of the species

Family *Hyaenidae* Gray, 1821.

Genus *Pachycrocuta Kretzoi*:

- Species *Pachycrocuta pyrenaica* (Depéret);
- Species *Pachycrocuta perrieri* (Croizet and Jobert);
- Species *Pachycrocuta brevirostris* (Aymard).

Genus *Crocuta* Kaup, 1828:

- Species *Crocuta crocuta praesplaea* Schütt;
- Species *Crocuta crocuta petralonae* Kurten;
- Species *Crocuta crocuta intermedia* (de Serres);
- Species Crocuta *crocuta spelaea* (Goldfuss);
- Species *Crocuta crocuta* Erxleben, 1777 (spotted hyena).

This family is composed of four genera recognized in the Pliocene, the Plio-Pleistocene and the Pleistocene.

The genus *Pachycrocuta* Kretzoi, with a voluminous skull, is said to derive from the Miocene genus *Hyaenictitherium*.

The dental formula is as follows: 3/3 I, 1/1 C, 4/3 P, 1 M.

It is known from three species of the European Pleistocene. *Pachycrocuta pyrenaica* (Depéret), an ancestral form of *Pachycrocuta perrieri*, was identified in the Pliocene sites of Serrat d'en Vaquer, Layna, La Callera and Odessa (Cregut-Bounnoure, 1996). *Pachycrocuta perrieri* (Croizet and Jobert) was close to *Pachycrocuta brunnea* and may have a common ancestor (Kurten 1968). The species appeared from the Lower Villafranchian (Les Étouaires, France) to the Final Villafranchian. It was present in many sites in the French Massif Central and Italy (Saint-Vallier, Villaroya, Senèze, Le Coupet, Valdarno and Blassac). In the Middle Pleistocene, it was found in Petralona (Greece) and Mauer and Mosbach (Germany). The third species *Pachycrocuta brevirostris* (Aymard), stronger than the previous one, was a Eurasian taxon known in Europe from the Middle Pleistocene. It was found in Sainzelles and Vallonnet (France) and Venta Micena (Spain).

The genus *Crocuta* Kaup has less grinding dental characteristics than the genus *Hyaena* and has as ancestor the Villafranchian form *Crocuta sivalensis* Falconer and Cautley.

The dental formula is: 3/3 I, 1/1 C, 4/3 P, 1-0/1 M.

This taxon is known by four species with subspecific affinities. *Crocuta crocuta praesplaea* Schütt known from the Pleistocene sites of Mosbach, Stranska Skala, Pétralona, Forest-Bad and Nauterie. *Crocuta crocuta petralonae* Kurten derived from the previous species but with much stronger characteristics. It was found in the upper ground of Petralona. *Crocuta crocuta intermedia* (de Serres) presented premolars much thicker than the cave hyena. Modified several times, Kurten and Poulianos (1981) ruled on the new species as an immigrant form from Africa. It was found in the Middle and Upper Pleistocene French sites (Lunel-Viel, Nauterie, Chatillon-Saint-Jean and Fontéchevade). *Crocuta crocuta spelaea* (Goldfuss) was a cave hyena that had a wide geographical distribution and was found in Upper Pleistocene sites (Cregut-Bounnoure 1996).

According to WoldeGabriel *et al.* (2001), the genus *Crocuta* has been present in Africa from the final Miocene, notably in the Adu Asa formation of the Awash in Ethiopia, whereas the species *Crocuta crocuta* was present in South Africa (Magapansgat 4 and Sterkfontein 4) in the Upper Pliocene. Several South African subspecies were also described in the Lower Pleistocene, but do not warrant this subspecific distinction. *Crocuta spelaea capensis* and *Crocuta crocuta venustula* were found at Kromdraai A and *Crocuta crocuta ultra* at Sterkfontein 1.

The oldest mention of *Crocuta* in North Africa is described by Geraads in the famous and rich Moroccan Upper Pliocene deposit of Ahl al Oughlam as *Crocuta dbaa* . Its dental characters were characterized by a long lower first molar, a high lower third premolar and a significant development of the cingulum on the last premolar and first molar.

In Algeria, it was from the Lower Pleistocene of Ain Hanech that *Crocuta crocuta* made its appearance. Arambourg's study (1979) describes a mandible and some isolated teeth with thick premolars and molars with notably a great development of the carnassial with an absent metaconid. In the Upper Pleistocene, the sites in the Algiers region, on the other hand, offered more dental and bone remains. First of all the discovery in the last century of a complete skull found in the Sintes cave in Ain Benian demonstrates the massive cranial and dental characters of a very large sized species which recalls the speleological species of the European caves. Then the fauna from the Warthogs site, rich in carnivorous remains, including a good number of the bones of large herbivores, presented at their extremities the bite and cutting marks of this hyena (Hadjouis 1985a, 1994; Hadjouis and Roubet 2020). Likewise, the presence in this site of several perfectly preserved coprolite remains indicated regular use of the site. The very large size and the massiveness of the dental and skeletal material were identical to those of the individual found in the Sintes cave.

6.4. The striped hyena (*Hyaena hyaena*)

Family *Hyaenidae* Gray, 1821.

Genus *Hyaena* Brisson, 1762.

Species *Hyaena prisca* de Serres.

Species *Hyaena hyaena* Linnaeus, 1756 (striped hyena).

6.4.1. *Chronological, geographical and morphological indications of species*

The genus *Hyaena* Brisson is characterized by a post-cranial skull and skeleton smaller than those of *Crocuta*.

The dental formula is as follows: 3/3 I, 1/1 C, 4/3 P, 1/1 M.

One species is known as *Hyaena prisca* de Serres, identified from the Mindel in the Grotte de l'Escale, at Es-Taliens and Montsaunès and at Mindel-Riss in Lunel-Viel and Igue-des-Rameaux, and Riss-Würm in Verzé and Montmaurin (Cregut-Bounnoure 1996).

The genus *Chasmaporthetes* Hay is known in Europe only by the unique species *Chasmaporthetes lunensis*, present in the Villafranchian sites of the Massif Central (Perrier-Pradines, Saint-Vallier, Senèze, etc.).

The striped hyena, which is smaller than the spotted hyena, was described in East Africa (Schungura B 10 and Usno formation) as early as 3 million years ago, then in the Lower Pleistocene of Tanzania (Olduvai Bed II).

The oldest remains attributed to this taxon in North Africa were found in the Tunisian Lower Pliocene deposit of Ain Brimba (Arambourg 1979). This author describes *Hyaena striata precursor* on the basis of a series of isolated teeth that seemed to belong, according to him, to the same mandible. The diagnosis that he gave characterized a striped Hyena with more robust teeth than that of the living form, with more developed posterior premolar lobes and a protruding cingulum. One year after the post-mortem publication of Arambourg's monograph (1979), Howell and Petter compared this subspecies to *Pachycrocuta*, which could be *P. pyrenaica*, whose dental characteristics seemed to be identified with the Pyrenean species of Serrat d'en Vacquer described by Depéret in 1890.

In addition, several remains of a Hyena were described by Arambourg in 1959 in the Upper Miocene (Vallesian) deposits of Oued el Hammam in Bou Hanifia and Marceau in Algeria under the name of *Hyaena algeriensis*, renamed by Howell and Petter (1985) as *Allohyaena (Dinocrocuta) algeriensis*.

In the Upper Pleistocene, its presence was attested to Ain Benian, probably at Grand Rocher near Algiers and Djebel Roknia. These two mentions were cited by Vaufrey (1955).

6.4.2. *Occlusal posture and postural balance of hyenas on the ground*

The spotted hyena and the striped hyena have a very characteristic body whose front limbs are higher than the hind limbs. Because of this structural adaptation, the withers are much higher than the hindquarters, with the back sloping downward, and the hindquarters very low (Figure 6.5). However, the unevenness of the limbs is somewhat different in the two species, where the dorsal inclination is replaced by an arched back in the striped hyena (*Hyaena hyaena*) (Figure 6.6). In contrast to some carnivores (*Felidae* and *Canidae*), the head is positioned lower than the withers and not in the horizontal extension of the cephalo-caudal axis.

But it is especially the cutting teeth, especially those of the spotted hyena (*Crocuta crocuta*), that are important in these species. The carnassial specialization (upper PM3 and lower M1) in longitudinal blade teeth has made their operation striking and efficient, acting like the blades of a pair of scissors (Figure 6.7). It is indeed in this species that carnassials play a major role in food laceration and constitute two symmetrical occlusion keys, together with the canines, ensuring the powerful locking of the jaws. The sharp blade tooth of the lower first molar and the sectorial trigone of the fourth premolar form a real shear, even before the canines intervene (Gaspard 1972).

Figure 6.5. *Posture of the head and quadruped body of the spotted hyena (Crocuta crocuta). The postural balance is broken by withers higher than the hindquarters, positioning the center of gravity at the front of the body, with the neck and head above the withers (© Hadjouis and Le Bihan)*

Figure 6.6. *Head and quadruped body posture of the striped hyena (Hyena hyena). In contrast to the spotted hyena, the arched back positions the head below the hindquarters and continues the arch shape (© Hadjouis and Le Bihan)*

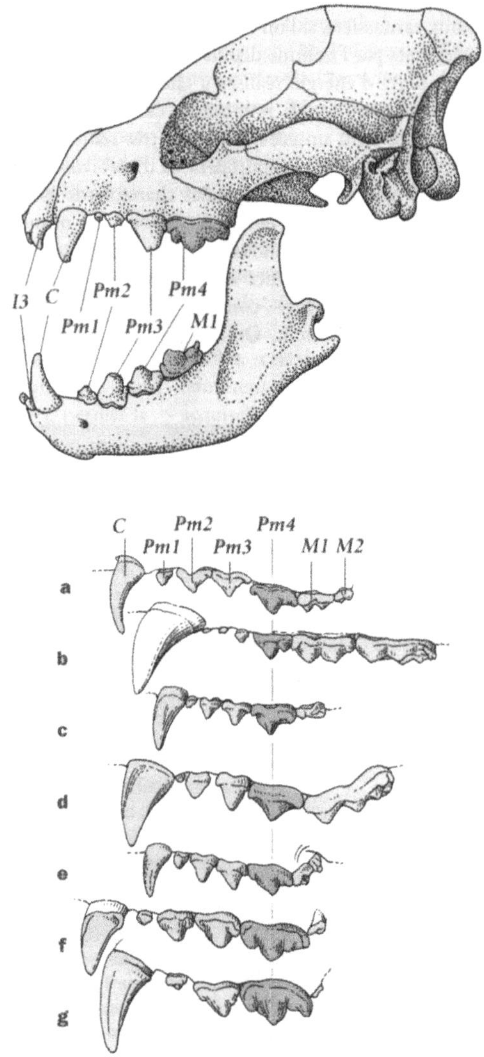

Figure 6.7. *Spotted hyena skull (Crocuta crocuta) in left lateral view showing the upper P4 and the lower M1 called carnassials and functioning like the cutting blades of a pair of scissors and comparison with the upper carnivores of some carnivores (© Éditions Belin/Dominique Visset).*

COMMENT ON FIGURE 6.7. – *a) Wolf (Canis lupus), b) bear (Ursus), c) pine marten (Martes martes), d) badger (Meles meleS), e) mongoose (Herpestes), f) hyena (Hyaena hyaena), g) lion (Panthera leo).*

6.5. The cave bear (*Ursus spelaeus*) and the brown bear (*Ursus arctos*): chronological, geographical and morphological indications of the species

Family *Ursidae* Gray, 1821.

Genus *Ursus* Linnaeus, 1758.

Species *Ursus spelaeus* Rosenmüller and Heinroth, 1794.

Species *Ursus arctos* Linnaeus, 1758.

Appearing in the Miocene, the family *Ursidae*, known by the genus *Ursavus*, groups together large carnivores of the plantigrade type. Dentition is characterized by the reduction or disappearance of the anterior premolars, while the molars have a more specialized development in bunodont-type grinding. The canines are very powerful. As a general rule, the posterior molars of *Ursidae* cannot be confused with the teeth of other carnivorous species because of their anterior-posterior lengthening related to the adaptation of the vegetarian diet.

The complete dental formula is as follows: 3/3 I, 1/1 C, 4/4 Pm, 2/3 M. In some specialized species (*Ursus spelaeus*) the anterior premolars disappeared.

The body of this species reached considerable proportions. Both massive and powerful, the cave bear was characterized by a skull with marked frontal bumps, a complication of the relief of the occlusal surface of the molars, the front limbs were more elongated than the hind limbs giving a distinctly lowered posture to the hindquarters, and the tibia had a diaphyseal torsion (Argant 1991, 1996a).

The genus *Ursus* Linnaeus is represented by seven fossil species: *U. minimus* Devèze and Bouillet, *U. etruscus* Cuvier, *U. deningeri* von Reichenau, *U. spelaeus* Rosenmüller and Heinroth, *U. arctos* Linnaeus, *U. thibetanus* Cuvier and *U. maritimus* Phipps. Among these forms, the last one was only known in the late glacial stage, while the cave bear (*U. spelaeus*) remains the most frequent species in karst sites, due to its hibernation. Its bony remains have been found and collected by numerous speleologists exploring the karst beds of the Jura and Alps between France and Switzerland, and some geological laboratories, such as the musée cantonal de géologie de Lausanne, hold dozens of skeletons from the Mialet cave in the Gard (Figure 6.8). The prehistoric and paleontological sites found in all of Europe between the end of the Middle Pleistocene and the entire Upper Pleistocene show an impressive number of bony remains of cave bears.

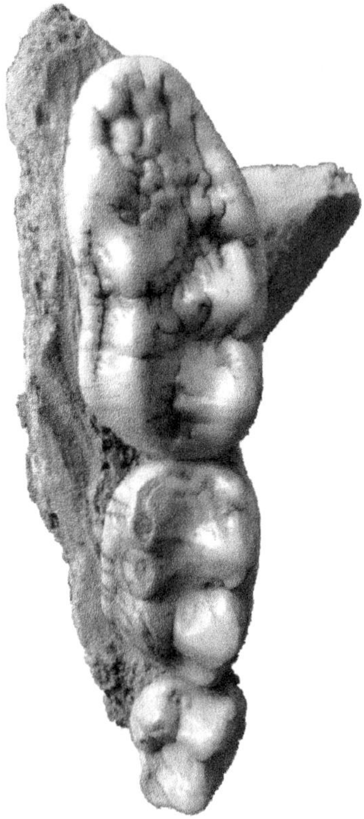

Figure 6.8. *Right hemi-jaw seen from above of cave bears (Ursus spelaeus) bearing P4-M2 from the Mialet cave (Gard) (© Hadjouis)*

French sites such as Fontéchevade, Chatillon-Saint-Jean, Cèdres cave, the Niche in Montmaurin or Azé have delivered the remains of both *Ursidae* species: *U. deningeri*, which reached the evolutionary form of *spelaeus*, and which will spread throughout the last glacial period and even beyond (Argant 1991).

The brown bear (*Ursus arctos* Linnaeus) is also large but less massive than the cave bear. The presence of the anterior premolars allows a specific distinction with *U. spelaeus*. The skull is elongated and demonstrates with the post-cranial skeleton a strong sexual dimorphism. The species, present throughout Eurasia during the Pleistocene, is probably of Asian origin (Shukutian in China) (Argant 1991).

The presence of large and small *Ursidae* was known in the Maghreb from the Pliocene to historical periods in numerous archeological and natural sites (Geraads

1997; Hamdine *et al.* 1998; Ouchaou and Amani 2002; Michaux and Thevenot 2007; Fontugne *et al.* 2012).

6.6. The wolf (*Canis lupus*): chronological, geographical and morphological indications of the species

Family *Canidae* Gray, 1821.

Genus *Canis* Linnaeus, 1758.

Species *Canis lupus* Linnaeus, 1758 (wolf).

The Family *Canidae* includes four genera. The genus *Canis* with the species *lupus* which will be described here and three other species (*Canis etruscus* Major, *Canis falconeri* Major and *Canis arnensis* Del Campana). The genus *Cynotherium* Studiati is represented by only one species *Cynotherium sardous* Studiati. The genera *Vulpes* Frisch and *Alopex* Kaup, small foxes, are represented by *Alopex lagopus* Linnaeus, *Vulpes praecorsac* Kormos and *Vulpes vulpes* Linnaeus. The last genus *Nyctereutes* Temminck is represented by the species *Nyctereutes magamastoides*.

Although of Miocene appearance, it was during the Villafranchian that the genus developed with a wide geographical distribution throughout Western Europe from a small species *C. etruscus*, which in all likelihood it was the ancestor of the present-day wolf.

The dental formula is as follows: 3/3 I, 1/1 C, 4/4 Pm, 2/3 M.

From the beginning of the Middle Pleistocene, the presence of the wolf was attested to in many European sites (Vallonnet, Escale, Caune de l'Arago, Mauer, Mosbach, etc.), even if sometimes the systematic position of the *Canidae* was subject to confusion because it is represented by two subspecific names: *Canis etruscus mosbachensis* Soergel and *Canis lupus mosbachensis*.

From the Middle Pleistocene onward, real wolves appeared throughout Western Europe, such as *Canis lupus lunellensis* Bonifay, whose stature was markedly larger (Cregut-Bounnoure 1996). The teeth were longer and thicker, especially on the M1, which had a high paraconid.

In the Lazaret Cave (Alpes Maritimes), a small Risian wolf was studied and compared with wolves from the south of France and Liguria (Hadjouis 1982).

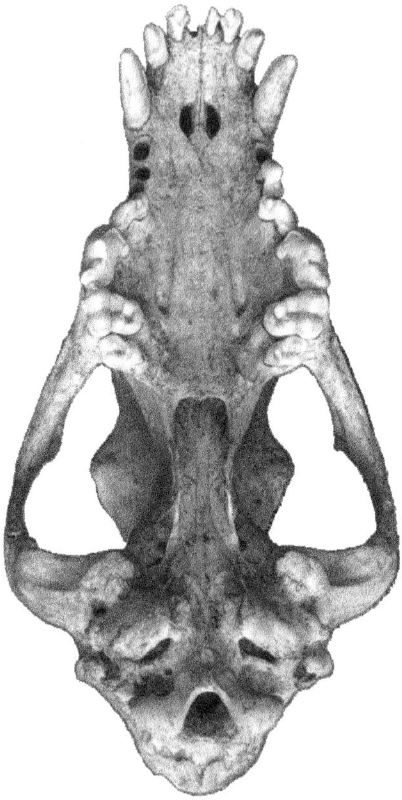

Figure 6.9. *Dental and craniofacial equilibrium in dogs seen from below. The two rows of incisors, canines, premolars and molars are aligned in the mesio-distal direction, as are the even parts of the arches and zygomatic processes, the articular tubercles of the temporal bone, the tympanic bullae, etc., are balanced and do not show asymmetries (© Hadjouis)*

Seen from below, the basicranial and palatal structure is perfectly balanced in *Canidae*. In both wolves and dogs, the incisor, canine, premolar and molar rows of the half jaws place the antagonistic rows in perfect balance without the slightest asymmetry (Figure 6.9). Similarly, from above, the craniofacial paired parts of the carnivorous quadrupeds are balanced and show the structural and occlusal harmony of these fossil canids of the Mediterranean Quaternary (Figure 6.10). The geometrical basion-prosthion triangle reflects the same extension of the craniodental and craniospinal fields (Figure 6.11). The perpendicular line of the triangle passes through M2 and well before the temporal line, which is not the case for *Felidae* with a reduced face size.

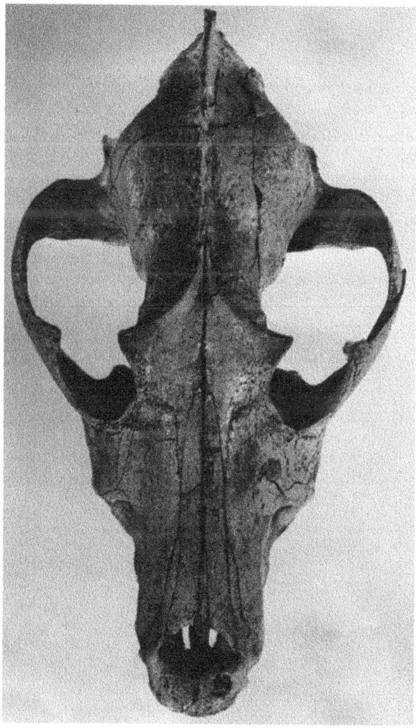

Figure 6.10. *Wolf skull (Canis lupus, S11.484) seen from above, from the Lazaret cave (Maritime Alps). The symmetry of the craniofacial paired parts of carnivorous quadrupeds is perfectly noted in this Mediterranean Quaternary canid (temporal fossae, frontal humps, zygomatic processes of the frontal and temporal bones, nasal bone, etc.) (© Kaufmann)*

Figure 6.11. *Wolf skull (Canis lupus, Q11.7) in left lateral view, from the Lazaret cave (Alpes maritimes). The geometric basion-prosthion triangle shows the extension of the craniodental and craniospinal fields. The perpendicular line of the triangle passes through M2 (© Kaufmann)*

7

Lagomorphs: The Hare (*Lepus capensis*)

Order *Lagomorpha* Brandt, 1855.

Family *Leporidae* Fischer, 1817.

Genus *Lepus* Linnaeus, 1758.

Species *Lepus capensis* Linnaeus, 1758 (hare).

7.1. Chronological, geographical and morphological indications of the species

Although they resemble the species of the Order of Rodents, particularly by their large, curved and powerful incisors, they do not have any particular connection with the latter. Lagomorphs have one pair of incisors per half jaw at the maxillary, whereas only one incisor grows in Rodents. A long diastema separates the incisors from the rest of the jugal teeth. The canines are absent. The absence of posterior closure of the glenoid fossa, which is not laterally delimited, allows lateral movements of the mandible in Lagomorphs thanks to special muscles. The wide latitude of the movements of the condyles then activates transverse rather than sagittal dynamics (Lecointre and Le Guyader 2001).

In ventral view, the incisive foramen, even orifices, are very developed and open posteriorly on the palate. In the same way, the maxilla and premaxilla also contribute to the palate.

The craniofacial structural model of the rabbit (*Oryctolagus cuniculus*) (same model as the hare) presents a large basic cranial flexion, which lowers both the orientation of the *foramen magnum* and the vault. The latter is parallel to the basi-occipito-sphenoidal floor (Figure 7.1).

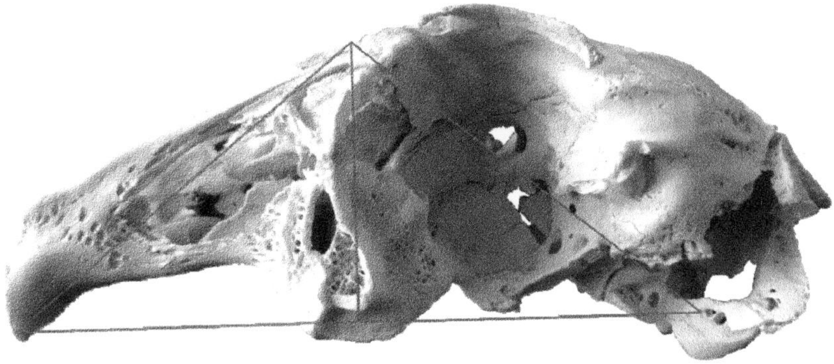

Figure 7.1. *Rabbit skull (Oryctolagus cuniculus) in left lateral view. The triangle shows a strong basic cranial flexion (© Hadjouis)*

Figure 7.2. *Head and body quadruped posture of the cape hare (Lepus capensis). The postural balance is broken due to an excessive elongation of the hind limbs, and it finds itself more often than not in a sitting rather than standing position. In reality, the animal if not running is sitting, the head is positioned above the withers (© Hadjouis and Le Bihan)*

The hare has developed hind legs that allow it to jump and run very fast. It is a quadruped digitigrade or plantigrade (Figure 7.2). The forelimbs are never prehensile as they can be in Rodents. The hare lives in fields, meadows and small woods. The species was first described in the Cape of South Africa and its range covers the Old and New World.

During the zoological exploration of Algeria in the 19th century, authors described several local breeds of hares. Seurat (1930) mentions that until 1898 Barbary hares were reported to be either *Lepus mediterraneus* Wagner or *Lepus aegyptius* Desmaret. He clarified that in reality the term Mediterranean hare was designated for the hare of Sardinia, different from the African variety and that the hare of Egypt was one of Lower Egypt, also different from the forms found elsewhere in North Africa.

From 1898, a large number of small hare forms were described in Algeria. Winton described the hare of Tell *Lepus kabylicus*, then *Lepus atlanticus*; Barrett-Hamilton, that of the Highlands *Lepus pallidior*, of which specimens were found at the Sfa Pass near Biskra. In 1902, Thomas Whitakeri named the small hare of the Sahara, *Lepus whitakeri*, finally another breed was found in Ain Sefra, buff-colored, described under the name of *Lepus sefranus*. The works of Angel Cabrera of the Museum of Madrid also describe, notably in his work *Les Mammifères de la faune ibérique*, several species of Berber hare.

Today, the Algerian varieties recognized as subspecies are those of: *Lepus capensis atlanticus* Winton, 1898 and *Lepus capensis whitakeri* Thomas, 1902. In the Quaternary, the bony remains of small hare were extremely rare in Algeria.

7.2. The status of the hare over the centuries

A well-known species in terms of its presence in number of remains found in archeological excavations; the hare remains known nevertheless in spite of bony remains of rabbits whose phenomenal fertility was demonstrated in tapestries from the 15th and 16th centuries. The medieval nobility had on the other hand a preference for hare hunting, as well as hunting for roe deer, collecting antlers, even modest in size, and a greater mass of meat (Audoin-Rouzeau 1994). French coats of arms and sculpture show a preference for the hare, while the rabbit belongs more to Genre paintings (Audoin-Rouzeau 1994).

PART 2

The Skull of Fossil Bipedal Vertebrates: Craniofacial Structure and Postural Balance

8

Primates

Although they are considered to be quadrupeds, Primates are voluntarily included in the chapters devoted to bipeds because of the phylogenetic relationship of certain groups whose hominoid lineages joined that of the *Hominidae* as early as the Upper Miocene.

The Order of Primates today includes the Super-Family of *Hominoidea*, formerly constituted by the Sub-Orders of Monkeys, Tarsier and Lemurs, called Old World Monkeys (*Catarrhini*) and New World Monkeys (*Platyrrhini*). The *Catarrhini* (thin nasal septum) populate Africa and Asia, while the *Platyrrhini* (wide nasal septum) live from Mexico to the north of Argentina. Adapted to arboreal life and plantigraphy, they have five well-developed fingers, with a separation of the big toe and toe from the other digits, used for grasping.

The Family *Hominidae* was separated from the Family *Pongidae*, tailless primates, which includes gorillas (*Gorilla*), orangutans (*Pongo*) and chimpanzees (*Pan*). New phylogenetic classifications and modern genetic analyses have shown that the evolutionary divergence tree did not correspond to this initial pattern.

Today, the Family of *Hominidae* becoming the sibling group of the Family *Hylobatidae* is composed of genera and fossil species with present-day humans (*Homo sapiens*), including the species of the three genera of Pongids, with however changes in the nomenclature of sub-families and tribes, as mentioned below.

> Super-Family *Hominoidea*.
>
> Family *Hominidae*.
>
> Sub-Family *Homininae*.

Tribe *Hominini*.

Genus *Homo*, Pan.

Super-Family *Hominoidea*.

Family *Hylobatidae*.

Sub-Family *Ponginae*.

Tribe *Gorillini*.

Genus *Gorilla*, Pongo, Hylobates.

The Family *Hominidae* has similar morphofunctional characteristics. The dental formula of *Catarrhini* is the same as in humans: 2 I/2, 1/1 C, 2PM/2PM, 3M/3M, which corresponds to 32 teeth, while *Platyrrhini* have 36 teeth. The loss of a premolar was already acquired in *Catopithecus* 36 My, indicating a *Catarrhini*/*Platyrrhini* discrepancy well before this period. The development of canines is reminiscent of that of Carnivores.

8.1. Occlusal posture, quadrupedic and verticalization of the Hominoid body

In terms of postural occlusion and adaptation to the ground, Primates react like quadrupeds and not like bipeds, even if they do so occasionally and for a short period of time. Functional musculoskeletal repercussions, whether normal or abnormal, are also found in the same postural dynamics as quadrupeds. However, from an early age, Great Apes adapt to the ground according to a bipedal postural balance, while having an axial growth and a Great Anthropoid occlusion.

The occlusal posture of the Miocene Hominoids was on the whole modeled on that of Great Apes: a face projected forward, reflecting the great antero-posterior development of the palate and mandible, a U-shape of the maxilla and mandible, the presence of diastema (lateral incisor space/canine space at the top and canine space/premolar space at the bottom) allowing the powerful canines to be housed there when closing the mouth and locking the occlusion, the large size of the teeth, a strong craniofacial musculature. Occlusal dynamics also responded to the craniofacial dynamics, including an adaptation of the latter with its occipito-cervical articulation. Even in the absence of a spine and pelvic girdle, the base of the skull alone of these Hominoids shows that it responded to a quadruped adaptation, thanks to a spine in the same configuration as that of Great Apes, i.e. a unique curvature with posterior convexity on a pelvis that was clearly higher than it was wide and intimately linked to a pelvic retroversion dynamic. Retroversion in fact led to maintaining the lumbosacral region straight or kyphosis, making it impossible to straighten the spine. The reduction of the posterior limb was in relation to a *Genu varum* and knee flexion. The postural functions of the latter imparted on the tibial compartments a fairly pronounced posterior tibial slope. The anterior limb was long. The gravity slope followed neither a true cephalocaudal axis in quadrupeds, nor a cephalopodal axis in humans, but rather an oblique cephalocaudal axis due to the significant elongation of the anterior limbs. The center of gravity was positioned at the level of the thoracic vertebrae.

The geometric outline of structural mechanics translates a skull-face relationship (anterior point of the occipital hole [basion] and anterior point of the dental alveoli [Prosthion] are quite comparable between Hominoids and Great Apes). The skull in extension was characterized by an absence of basi-occipital flexion, a not very verticalized clivus whose sphenoidal angle remained excessively open, a craniofacial field slightly more developed antero-posteriorly than the craniospinal field, at times equivalent.... The angle of the upper triangle remained open by more than 75° reflecting the absence of a greater verticality in the slope of the clivus (Figure 8.1).

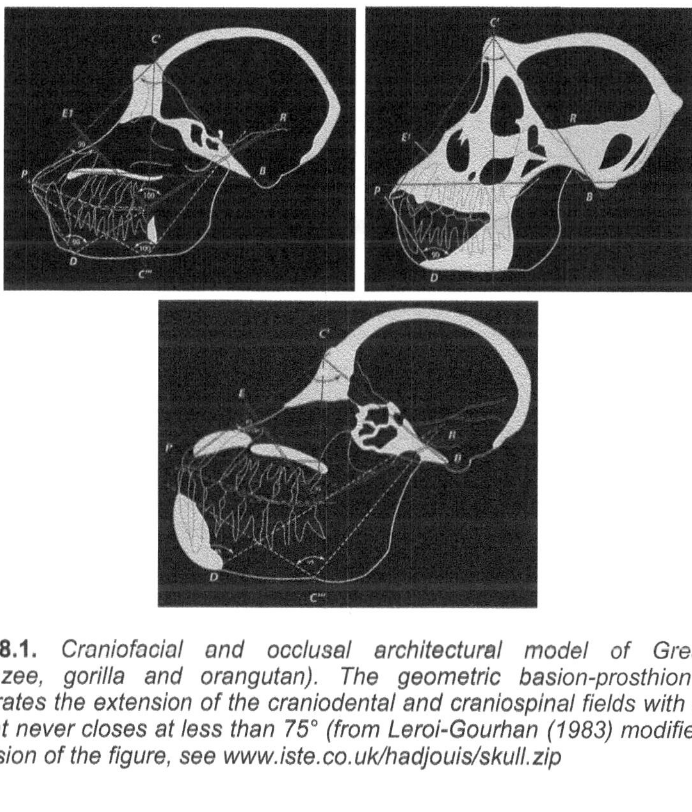

Figure 8.1. *Craniofacial and occlusal architectural model of Great Apes (chimpanzee, gorilla and orangutan). The geometric basion-prosthion triangle demonstrates the extension of the craniodental and craniospinal fields with an upper angle that never closes at less than 75° (from Leroi-Gourhan (1983) modified). For a color version of the figure, see www.iste.co.uk/hadjouis/skull.zip*

Given cuspid interlocking constantly maintaining the dental articulation in labidontia, following a craniofacial structure whose posture itself followed a logic of quadruped equilibrium, the face (or the whole craniofacial complex), like all herbivorous, omnivorous and carnivorous Mammals, was balanced, presenting neither asymmetry nor malocclusion. In the end, cuspid interlocking also contributed to maintaining the lower and middle levels of the face in perfect facial harmony.

8.2. Work in dentofacial orthopedics and embryogenesis

FDO work, and in particular that of Delattre and Fenart (1952, 1970), highlights by the vestibular method, a rotation and tilting of the squamous part of the occipital bone around a vestibular axis (axis of the semicircular canals of the inner ear). Deshayes, like Dambricourt Malassé, believes that the early embryonic stages are crucial to understanding the complexification of basic cranial flexion. It is in the mesoderm that morphogenesis takes on its full significance in the early intrauterine period. The mesoderm is subjected to direct constraints (attachments of the dura membranes to the basi-sphenoid and basi-occipital) or indirect constraints (attachments or tendons on the anterior and posterior clinoid apophyses of the sella turcica) in relation to modifications of the neural tube (Deshayes 1988). Moreover, Melsen (1974) and Deshayes (1986) both demonstrate the importance of activity at the level of spheno-occipital synchondrosis (clockwise rotation of the sphenoid body, counterclockwise rotation of the occiput, descent of the squamous part of the occipital bone below the craniopalatine line, and forward projection of the occipital condyles, frontalization of the petrous pyramids, etc.).

For Dambricourt Malassé, craniofacial growth and in particular occipital tilting must be studied above all by their ontogeny because the study of the embryonic and fetal periods is as important as that of the post-natal periods (1992–1996). This is confirmed in comparative work with Great Apes in the juvenile and adult growth stages, particularly with the difference in endocranial axial verticality gain quantified by the angle formed between the sphenoidal plane and the spheno-basilar slope. The relative position of the homologous points at the base of the skull changes rapidly by opening the sphenoidal angle and moving up the occipital scale. The differences noted between these two dynamics are already acquired during intrauterine development (Dambricourt Malassé 2011). Likewise, the craniofacial and craniospinal fields are equivalent, but the craniopalatal axis of the Great Ape does not tilt forward and downward as a human fetus would. After birth, there is a lengthening of the craniofacial field and a disappearance of the craniopalatine balance, losing forever the possibilities of bipedal posture.

In other words, the axial skeleton of the three genera *Gorilla*, *Pan* and *Pongo* remains unchanged in terms of their organization in space: the same degree of verticalization starting with a plicature of the base, same growth vectors.... However, and taking into account these ontogenic observations, Dambricourt Malassé (2011) insists on the fact that the craniofacial morphogenesis of current Great Ape species should be a "model" allowing the reconstitution of great lineages of embryonic morphogenesis of fossil Great Ape species and not the model for the oldest *Hominidae*.

9

Hominoids

The comparative anatomy of Primates and that of fossil *Hominidae* has been studied and taught in recent years by some paleontologists. Senut has not hesitated to criticize the way studies were devoted to the origin of humans by including classical comparisons of *Australopithecus* with chimpanzees and present-day humans. Comparative ape traits were considered archaic or primitive, yet it appears from molecular and anatomical analyses of Great Apes that their traits are rather derivative (Senut 2005).

While the oldest Hominoids came from the Upper Oligocene (between 27.5 My and 24.2 My) of the Lothiodok Hills (Kenya) known as *Kamoyapithecus hamiltoni*, it was in the Lower Miocene (between 21 My and 17.5 My) that the diversification of the Great Apes was significant in Africa. The genus *Proconsul* Hopwood, discovered in 1928 at the Koru site, is the best known, notably through the description of several species (*P. nyanzae*, *P. africanus* and *P. heseloni*) (Hopwood 1933; Le Gros Clark and Leakey 1950; Walker and Pickford 1983; Walker *et al.* 1993). Sexual dimorphism was very marked in these species, some of which (*P. heseloni*) may have represented females.

Among the most striking anatomical features were: strong upper transverse mandibular torus, absence or weak development of the lower transverse torus, gracefulness of the mandibular body, weak or even absent supraorbital torus, strong zygomatic bone, pronounced muzzle, fine enamel of premolars and molars, enlarged upper central incisors, the lower ones being tall and narrow, canines with strong dimorphism, upper molars with large hypocone, with lingual cingulum protruding from the upper molars, lower molars with wide trigonid surrounded by five prismatic cuspids and a large hypoconulid.

The forelimbs were as long as the hind limbs and their locomotion suggests an arboreal quadruped and occasionally bipedal walk (Walker and Pickford 1983; Senut 1989; Harrison 2002).

In the most recent Lower Miocene levels (16.5 My) in Moroto (Uganda), the discovery of new bone and dental remains made it possible to change the status of *Proconsul major* into a new genus called *Ougandapithecus* based on canine morphology (Senut *et al.* 2000). The new characteristics of *Ougandapithecus major* were: a very large size, close to the gorilla, marked sexual dimorphism, canine at the top in laminaria, very large molars with low tubercles, strong upper mandibular torus, non-existent inferior, femur with a small trochanter not protruding much on its posterior face, a large trochanter located below the femoral head and tibia with medio-lateral compression.

In the Middle Miocene (between 17 My and 11 My), Great Apes were widely distributed on the former Arab-African continent (East Africa, Southern Africa and Saudi Arabia). Bone and dental remains were discovered in Saudi Arabia (Ad Dabiyah) and Uganda, but it was in East Africa that the Great Apes were most numerous: Maboko, Fort Ternan, Ngororo in Kenya, Moroto in Uganda, and represented by the genera *Afropithecus*, *Kenyapithecus*, *Nacholapithecus* and the Namibian genus *Otavipithecus*.

Afropithecus was characterized by: a strong upper transverse torus of the mandible, a maxilla with a long, narrow and protruding muzzle, a well-developed maxillary and ethmoidal sinus with an enlarged anterior region, very prognathous muzzle, shallow palate, wide interorbital space, high canines with circular section, thick dental enamel, very wide upper premolars, protrusive and robust central incisors and very reduced laterals (Senut 2005).

9.1. *Kenyapithecus*

Known by two species: *Kenyapithecus wickeri* and *Kenyapithecus africanus* with a robust morphology. Mandibular and dental characters were characterized by: a robust mandible, strong lower transverse torus, weak upper, a proclive mandibular symphysis, vertical mandibular ramus starts at M2, broad premolars, thickened tooth enamel and maxillary M3 larger than M2. The skeleton of the limbs indicated a probably mixed locomotion, suggesting arboreal and terrestrial quadrupedics (Senut 1989).

9.2. *Nacholapithecus*

Represented by several skeletons discovered since 1982 in the Nachola deposit (Kenya) (Ishida *et al.* 1984; Sawada *et al.* 1997; Nakatsukasa *et al.* 1998). These remains suggest a large Primate whose dimorphism appeared significant, the base of the zygomatic arch was low, placed above the M1-M2, canine fossa, sub-nasal region and palate weakly elongated, mandibular body high and not very thick, low and robust crown of the canines and anterior limb longer than the posterior.

9.3. *Otavipithecus namibiensis*

The species whose cranial, dental and post-cranial remains found in 1991 in fossiliferous breccias in northern Namibia (Conroy *et al.* 1992; Pickford *et al.* 1994; Senut and Gommery 1997) suggest a large Primate. Morphological characteristics include: a frontal with large frontal sinuses, weak supracilliary arches and an absent supraorbital torus. The mandible had the same height of the mandibular ramus over its entire length, a narrow incisive region, a large retromolar space and thin tooth enamel (Thomas and Senut 1999).

10

From Hominoids to Hominids

The Upper Miocene period (between 11 My and 5.5 My) is interesting in two ways because environmental changes were conducive to the radiation of *Pongidae* in southern Eurasia on the one hand, and on the other hand the end of the Miocene-Pliocene transition saw the development of *Australopithecus* in Africa. The representatives of this period are the genera *Samburupithecus*, *Ardipithecus*, *Orrorin* and *Sahelanthropus*.

The jawbone fragment recovered by Ishida and her team in 1982 in Kenya (Ishida *et al.* 1984; Sawada *et al.* 1997) enabled her to describe a new genus and species of primate called *Samburupithecus kiptalami* (Ishida and Pickford 1998; Pickford and Ishida 1998). The most significant morphological features of this gorilla-like species were: low location of the root of the zygomatic arch above M2, a deep arched palate, thickening of the tooth enamel and swelling of the molar cusps and decrease in molar size from M3 to M1 (Thomas and Senut 1999; Senut 2005).

10.1. *Ardipithecus ramidus kadabba*

This is a subspecies that originated in Ethiopia in 2001 following the discovery of a few fossil remains (Hailé-Sélassié 2001; WoldeGabriel *et al.* 2001). The few isolated teeth and post-cranial fragments suggest clear differences with the current Great Apes, particularly in the morphology of the canines in the form of incisors. On the other hand, the characteristics of the skeleton of the limbs are reminiscent of those of Great Apes, suggesting a locomotion that was still arboreal.

10.2. *Praeanthropus tugenensis* (= *Orrorin tugenensis*)

This is the species name given to the genus *Orrorin* whose remains, dated between 5.7 My and 6.1 My, were found in the Lukeino formation sites (Lake Baringo Basin, Kenya) in 2001 (Pickford and Senut 2001; Senut *et al.* 2001; Sawada *et al.* 2002). The remains that constitute this taxon, hominoid for some and hominid for others, consist of portions of mandibulars, isolated teeth and fragmentary remains of the skeletal remains of the limbs (proximal fragments of the femur, distal diaphysis of the humerus, proximal phalanx and distal phalanx of the first ray).

Dental characters bring them closer to chimpanzees and modern humans (smallness of the teeth), others of the current Great Apes and fossils (*Kenyapithecus*, *Proconsul*) like the morphology of canines and premolars. The teeth were also distinguished by a right mandibular symphysis, a thickening of the dental enamel of the molars and the absence of diastema (Senut 2005). The latter characterizing a clear reduction in the height of the canines.

The characters of the appendicular skeleton, obviously derivative, were found in Australopithecus and humans, denoting bipedal locomotion (long femoral neck, antero-posterior flattening of the femoral neck, anterior torsion of the femoral head, position of the lesser trochanter, etc.). However, the morphology of the humerus and phalanxes rather suggesting an arboreal adaptation (Gommery and Senut 2002).

10.3. *Sahelanthropus tchadensis*

While for a very long time paleontological research was concentrated in East Africa, then in South Africa, nobody suspected the presence of Hominoid and Hominid fossils in sub-Saharan Africa, in the middle of the Sahelian territory. For more than 20 years, Brunet and his Franco-Chadian team criss-crossed Chad, first in search of Miocene Large Mammals. It was in 2002 that his first Hominoid discovery was made around the world. The new taxa of the genus and species *Sahelanthropus tchadensis*, about 7 My, are essentially based on the presence of a partially crushed skull (Vignaud *et al.* 2002). The inventors placed the specimen, not in the lineage of the Great Apes, but in the origin of the Hominids, making this taxon the oldest of the lineage. Their arguments were based on the small canine tooth, flattening of the face, anterior position of the *foramen magnum*, robust morphology probably related to dimorphism and crowns of the lower jugular teeth (Brunet *et al.* 2002). However, the paleontologist community is divided: first on the reconstitution of the calvarium, and then on the arguments put forward defending the hypothesis of a Hominid rather than a Great Ape. The flattened face of the specimen, identified as male, was in contradiction with the smallness of the canine (crown and root). The root development in males considerably increased the superior and anterior-apical region, giving the face a bulging morphology, which was not the case here. As for the virtual reconstitution of the fossil, the parameters of analysis seemed erroneous, for example the points of the FM (fronto-mastoid) line passing through the top of the mastoid apophysis when it should pass through the basion. Here the line appears much less verticalized and passed being close to modern man, whose flexion is more advanced than in Australopithecus.

Moreover, the structural dynamics of the Chadian fossil is in a perspective belonging to the extensive craniofacial dynamics of quadrupeds. In other words, the flattening of the nuchal floor, the absence of basi-occipital flexion, the great anterior-posterior extension of the neurocranium, even greater than that of the Australopithecus, and the great height of the crown of the canine (despite its smallness) exceeding the occlusal plane, thus inscribing its labidontal occlusion with cuspid interlocking, all pointing in the direction of the craniofacial and occlusal dynamics of Great Apes.

Other parameters indicate a need for more caution. Indeed, Dambricourt-Malassé, in her comparative work (2011), insists precisely on the size and occlusion of the canine. The latter phenomenon is comparable to a deciduous occlusion of a gorilla or chimpanzee with a similar orientation of the base. Similarly, the mechanisms of the base of the skull, the cornerstone of architectural understanding during phylogenesis, are not followed. Thus, she points out that if the cerebellum remained in an elevated position, the *foramen magnum* could not be lowered, as this was a consequence of the spatial remodeling of the other bones. However, the opposite is described. In agreement with other paleontologists, Dambricourt-Malassé refutes the idea that the Chadian fossil could be a good representative at the origin of the *Homindae* lineage:

> This species has never been verticalized beyond the embryonic rotation common to the three current genera of Great Apes, it remains in extension. It does not correspond to the structure of a Hominid imposing a permanent bipedal balance. (Dambricourt-Malassé 2011, author's translation)

10.4. *Ardipithecus ramidus*

It was in the Lower Pliocene of the Middle Awash Valley in Ethiopia that remains of numerous fossils of *Ardipithecus ramidus* were found in 1992 among faunal remains. The inventors consider this 4.4 My old taxon to be part of the *Hominidae* lineage (White *et al.* 1994; Wolde-Gabriel *et al.* 1994). Among the advanced characteristics: a *foramen magnum* in mesial position, oriented downward, the thickness of the dental enamel, the size of the canine tooth or the morphology of a proximal phalanx of the foot. However, several paleontologists do not seem to share the discoverers' point of view regarding the affiliation to *Hominidae*. Facial morphology and occlusal posture, close to *Sahelanthropus*, correspond to a cranial base that did not yet allow for bipedalism; these species still moved like *Proconsul* on rami balanced on four limbs (Dambricourt-Malassé 2011).

10.5. *Praeanthropus africanus* (= *Australopithecus anamensis*)

The species formerly known as *anamensis* was defined from fossil remains collected in 1994 in northern Kenya from two localities, Kanopoi and Allia Bay, dated 4.2 My and 3.9 My (Leakey *et al.* 1998). Several dento-skeletal remains were described, supplemented by skeletal remains of limbs found in 1966 by Patterson in Kanopoi (Patterson and Howells 1967). The inventors note some human morphological features such as the tibia or humerus.

11

Australopithecus

The genus and species of the oldest *Australopithecus* (*Australopithecus anamensis*), mentioned above, has undergone a taxonomic modification. The other more recent species are known between 3.5 My and 2.5 My in East and South Africa. Among the most representative taxa in terms of individuals are *Australopithecus afarensis* and *Australopithecus africanus*. For the first time, representatives of this lineage shared human and chimpanzee-like characters.

Since the first discovery attributed to the skull of the Taung child in South Africa, hundreds of fossil remains have multiplied in the two main African regions. The *Australopithecus* are grouped according to their cranial, dental and skeletal characteristics into two groups: one grouping *A. afarensis*, *A. africanus*, *Australopithecus bahrelghazali* and *Australopithecus garhi*; the other grouping *Australopithecus robustus*, *Australopithecus aethiopicus* and *Australopithecus boisei*. This second group is considered to include robust species, also called *Paranthropus*.

But doubt still persists among many authors to classify these three species in one or the other genus. However, it should be noted that these three taxa are distinguished from other more slender *Australopithecus* by a very marked sexual dimorphism where males developed a sagittal ridge, like that in gorillas, allowing the temporal muscles to activate their mandibles. *Paranthropus* also had strong muscles, strong teeth and a flattened face.

The classification of these *Hominidae* is presented below.

11.1. *Australopithecus afarensis*

Australopithecus afarensis Johanson, White and Coppens, 1978.

Holotype: mandible LH 4 (Laetoli, Tanzania).

Synonymies: *Meganthropus africanus* Senyürek, 1955; Pre-*Australopithecus* Coppens, 1981.

In addition to the classic species named "Lucy" whose sub-complete skeleton was found at the Hadar site (Ethiopia), the species is also known from several other cranial, mandibular and post-cranial remains of several juvenile and adult individuals, discovered in other localities in East Africa.

11.2. *Australopithecus africanus*

Australopithecus africanus Dart, 1925.

Holotype: Taung skull (South Africa).

Synonymies: *Australopithecus transvalensis* Broom, 1936; *Plesianthropus transvalensis* Broom and Schepers, 1946; *Australopithecus prometheus* Dart, 1948.

As *Australopithecus afarensis, A. africanus* is also known from many fossil remains, discovered in South Africa. Their remains are considered to be the oldest.

11.3. *Australopithecus bahrelghazali*

Australopithecus bahrelghazali Brunet *et al.*, 1995.

Holotype: mandible KT 12/H1 (Koro Toro, Chad).

This species, whose bony remains were found in 1995 for the first time in the Sahel, outside the classical territories of hominid fossil discoveries, yielded cranial and mandibular portions. The context of the discovery is a fluvio-lacustrine environment dated from 3 to 3.5 My. The mandible named Abel presents a mixture of apomorphic characters (derivatives) such as the molarization of premolars and plesiomorphic (archaic) characters such as the triradicular presence of premolars, like the Panins.

11.4. *Australopithecus garhi*

Australopithecus garhi Asfaw *et al.*, 1999.

Holotype: maxilla Bou-VP-12/130 (Bouri, Middle Awash, Ethiopia).

This species is contemporary of *A. aethiopicus* (2.5 My), but differed from it by the large size of the teeth and a prognathic face, although different from the prognathism of the robust species. The other characteristics were similar to those of the slender species.

In general, the slender species *A. afarensis*, *A. africanus*, *A. bahrelghazali* and *A. garhi* had a brain shape (350–500 cm^3) that differs from that of the chimpanzee, but the face was still prognathic. The base of the skull was elongated. The mandible had a high and thick mandibular ramus and the symphysis had a distal inclination of the chin. The presence of the maxillo-mandibular diastema (lateral incisor/canine space at the top and canine space/premolar space at the bottom) demonstrated the great development of the canine inserting itself between these spaces during the closing of the mouth, whose anterior-posterior dynamics considerably reduced movements by cuspid interlocking. The teeth were wide and the molars increased from the first to the third.

11.5. *Paranthropus robustus*

Paranthropus robustus Broom, 1928.

Holotype: skull TM 1517 (Kromdrai, South Africa).

Synonymies: *Paranthropus crassidens* Broom, 1949.

The two species of the genus *Paranthropus* (*P. robustus* and *P. boisei*) are known in later levels of the *Australopithecus* group. The species *P. robustus*, from South Africa, is dated at about 1.95 My.

11.6. *Australopithecus aethiopicus*

Australopithecus aethiopicus Arambourg and Coppens, 1967.

Holotype: mandible Omo 18-1967-18 (Shungura, Omo, Ethiopia).

Hominid remains were collected in levels C, D, F and H. The mandible was described by Arambourg and Coppens as first *Paraustralopithecus aethiopicus*, renamed *A. aethiopicus*. Dated between 2.6 My and 2.5 My, the massive shape of this mandible presented the characters of a robust but archaic individual as shown by the reduction of its incisal edge, the section of its symphysis and the extreme reduction of its alveolar planum. The new fossil evidence found since then shows that individuals of this *Australopithecus* species measured 1.50–1.60 m and had a brain capacity of 500–600 cm^3.

11.7. *Australopithecus boisei*

Australopithecus boisei Tobias, 1967.

Holotype: OH 5 skull (Olduvai, Tanzania).

Synonymies: *Zinjanthropus boisei* (Leakey, 1959); *Paranthropus boisei* (Robinson, 1960); *Australopithecus* (*Paranthropus*) *boisei* (Leakey, Tobias and Napier, 1964).

The robust *Australopithecus* or *Paranthropus* mentioned above present a reorganization of their craniofacial structure with notably a flattened face in *P. robustus* and *P. boisei*, a cerebral volume more significant than that of the slender *Australopithecus*, a cranial base more shortened and more flexed. On the other hand, some authors prefer to move the species *aethiopicus*, given the presence of primitive characteristics shared with the *Australopithecus* (prognathic face, flattening of the palate and the temporomandibular joint, etc.) and derived characters found essentially in *Paranthropus* and *Kenyanthropus* (very lateralized projection of the zygomatic arches, a flattened face in particular, alignment in the same plane of the incisors and canines, etc.) (Braga 2005).

Yet, several authors often forget that the diagnosis generally established in regards to the brain, the cranial capacity and the face, seriously neglect the base of the skull whose dynamic elements are directly linked to the cervical spine, which itself indicates its relationship with an arboreal and terrestrial or exclusively terrestrial adaptation. These features, such as basi-cranial flexion, spheno-occipital rotation, the frontalization of the petrous pyramids, the situation of spheno-occipital synchondrosis and the relationship of flexion with occlusion, are able to position a certain number of taxa. Indeed, the base of the KNM-ER-17000 fossil, not different from that of ARA-VP-1500, presents a spheno-occipital synchondrosis which attests to a much greater rotation of the body of the sphenoid than all previous fossils (Dambricourt-Malassé 2011).

12

The Genus *Homo*

Bipedalism was known, as mentioned above, with the *Hominidae* prior to the genus *Homo*. Although bipedalism is not exclusive to the genus *Homo* and that other Vertebrates share it like non-flying Birds or certain Saurians, we use this term, inscribed in the title of our series "Comparative Anatomy and Posture of Animals and Humans", to give the duality of quadrupedalism/bipedalism all its comparative significance in the implementation of new organizational plans. It is understood that the terminology of "verticality" or "straightening" of the body takes on a more logical meaning insofar as straightening led to hominization through the arrangement of a head above a verticalized spine. Thus, the whole importance of this series is that the mechanism leading to "bipedalism" or "verticalization" no longer followed the same quadrupedal or semi-bipedal centers of gravity. In the same way, the bipedal arrangement, as we know it today, is incomplete, full of anomalies and articular malformations, from the head with numerous occlusal shifts, to the feet whose adaptation to the ground is not yet balanced. And this postural adaptation is not new, since it started as soon as the body was straightened for a standing position.

The distribution of the centers of gravity in a bipedal station, as theoretically known in anatomical atlases (of the standing at attention type), is not in conformity with the structural and postural reality of the populations.

With the acquisition of bipedalism, significant repercussions occurred at the postural level, thus introducing new morpho-functional adaptations. The mechanical load is not distributed in the same way on the lower limbs and especially on the knee joint of fossil and current populations. In the latter, the load line corresponding to the gravitational points of the body is not uniform and responds to certain parameters related to the weight, bone structure, structural morphology, bipodal disposition and certain congenital, hereditary and familial pathologies of the individuals in a population.

The genus *Homo* was defined on the basis of the species *Homo habilis sensu lato* which was later split into two species: *Homo rudolfensis*, whose remains were found within the archeological material of Koobi Fora in Kenya, and *Homo habilis sensu stricto*, whose fossils come from Olduvai in Tanzania. Proponents of specific distinctiveness (Stringer 1986; Chamberlain and Wood 1987; Lieberman *et al.* 1988, 1996; Wood 1992; Rightmire 1993; Donnelly 1996; Prat 1997, 2000; Prat and Marchal 2005) point to a number of distinguishing characteristics, notably the variation in brain volume (510–674 cm^3 for *H. habilis* and 750 cm^3 for *H. rudolfensis*), sexual dimorphism, the degree of craniofacial biometric variation, the shape of the scale of the temporal bone, the width of the middle face, significantly greater in *H. rudolfensis*, a clear differentiation of the zygomatic bone and palate, etc.

The particularity of the discovery of *Homo habilis sensu lato* lies first of all in the association with its bone remains with tools carved in the same archeological levels. Then the comparison of its morphological, cranial, dentine and post-cranial characteristics defined by their discoverers, Louis and Mary Leakey, with the *Australopithecus* and *Homo erectus* allowed to rule on this species and to define it from a new diagnosis (Leakey *et al.* 1964).

Like the majority of cranial fossils, the base of the skull and its relationship with the cervical spine are rarely described. There are still several avenues to be explored in order to understand the relationship between the occipital tilt and spheno-occipital rotations on the one hand, and cranio-palatal, fronto-ethmoidal and finally occlusal rotations on the other hand. It is these relationships that make it possible to position the *foramen magnum* and the atlanto-occipital hinge. In other words, the sphenoidal flexion, the orientation of the cerebellum and the cranio-palatal equilibrium allowed Anne Dambricourt-Malassé, who has been studying these mechanisms applied to *Hominidae* for many years, to propose new models. These are based on organization planes, not on the craniofacial external relief (exoskeleton), but rather on the deep basic cranial structures, visible and analyzed in imaging. This methodology applied by orthodontists and FDO specialists, whose inspiration has reached renowned paleontologists and odontologists such as Jean Piveteau, Robert Gudin, André Leroi-Gourhan and Robert Fenart, allows us to understand the orientations of growth trajectories during phylogenesis. Anne Dambricourt-Malassé therefore insists, at the time of each new organization plan, on the problem centered on the lowering of the cerebellum and the concomitant modifications of the occlusion, with the dynamics of the sphenoid and its synchondroses as the articulation. The author insists once again on a feature that is clearly visible in imaging, namely, the basi-occipital plicature. This has been absent since the development of the first vertebrates to the first primates. The dynamic neuro-embryonic rotation was at the origin of axial verticalization, by means of a cerebellum amplifying its verticality at each evolutionary stage.

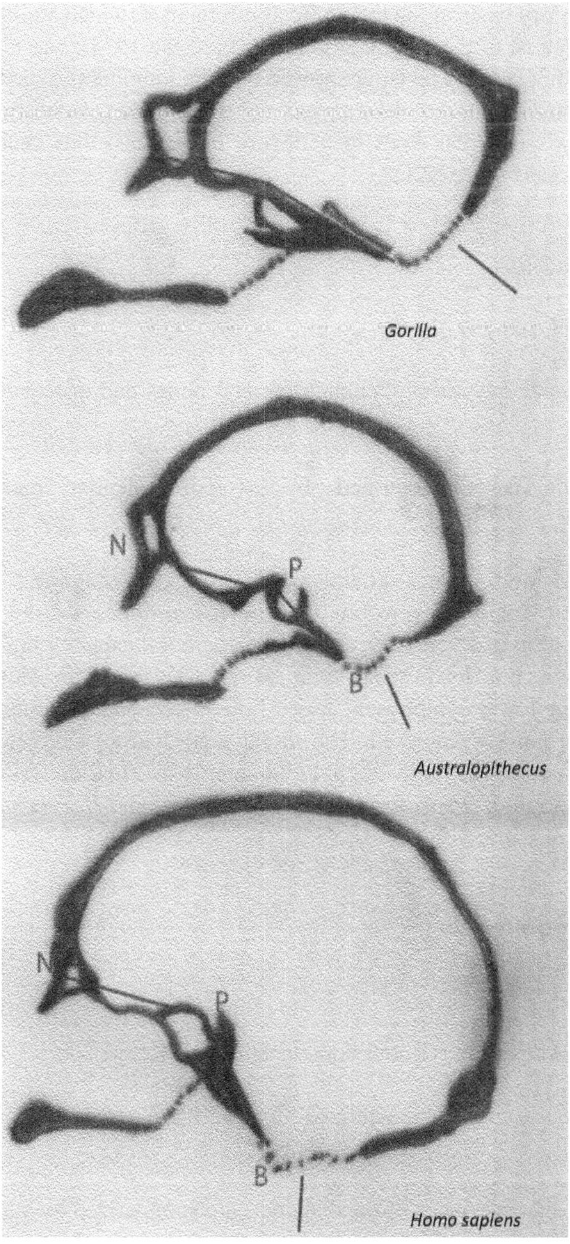

Figure 12.1. *Longitudinal sections of the base of the skull in the Gorilla, Australopithecus and modern man (after Le Gros Clark (1964)). For a color version of the figure, see www.iste.co.uk/hadjouis/skull.zip*

The explanatory basis of the craniofacial mechanism, which Vallois and Piveteau call "the cornerstone of the cranial structure" (Piveteau 1973), was already launched by Le Gros Clark (1964). The latter showed that the longitudinal sections of the base of the skull in the Gorilla, *Australopithecus* and modern man started from a rectilinear basic cranial axis in Great Apes to a flexure in humans, thus causing significant structural changes (Figure 12.1).

12.1. *Homo habilis*

Homo habilis Leakey *et al.*, 1964.

Holotype: several specimens including parietal bones and one mandible (Olduvai, Tanzania).

Homo habilis is characterized by a more slender morphology than *H. rudolfensis*.

The cranial and dental characteristics of the diagnosis are generally distinguished by smaller dimensions than in *Australopithecus* and fall within the interval of variation of *H. erectus*, as well as a cerebral capacity higher than that of *Australopithecus*, but lower than that of *H. erectus*. The chin was weakly developed, at the lower symphyseal edge of the lower mandible, and cranial superstructures were not very marked. The molar teeth had an elongated mesio-distal diameter, while the vestibulo-lingual diameter was smaller, falling within the interval of variation of *Australopithecus* (lower limit) and *H. erectus* (upper limit), the incisors were wide.

12.2. *Homo rudolfensis*

Homo rudolfensis Alexeev, 1978.

Holotype: KNM ER 1470 (Koobi Fora, locality 131, Kenya).

Homo rudolfensis is characterized by a more robust morphology than *H. habilis*.

The cranial characteristics of the diagnosis are summarized by very high orbits, a narrow nasal opening and a supra-orbital torus developed more medially than laterally.

12.3. *Homo ergaster* and *Homo erectus*

The first terminological designation of *Homo erectus sensu lato* was based on exclusively Asian fossil remains since the first discoveries from the end of the 19th century were made by Dubois in Trinil on the famous *Pithecanthropus*, followed by Zhoukoutian, then Java by Von Koenigswald.

It was between 2 My and 1.8 My that the oldest forms of *H. erectus* were found in East African sites at Olduvai and east of Lake Turkana. The most archaic African forms are called *Homo ergaster*. Other more typical forms with derived characteristics, *Homo erectus sensu stricto*, colonized much of Africa (South Africa around 1.5 My and North Africa around 1 My) as well as Eurasia around the same periods up to 500,000 years ago (Caucasus, Southern Europe, Middle East, China and Java) (Arambourg 1954, 1955; Arambourg and Hoffstetter 1954, 1955, 1963; Leakey 1976; Carbonell *et al.* 1995; Wanpo *et al.* 1995; Ascenzi *et al.* 1996; Rosas and Bermudez de Castro 1999; Gabunia *et al.* 2001; Vekua *et al.* 2002).

Homo ergaster Groves and Mazak, 1975.

Holotype: KNM ER 992 (Ileret, East of Lake Turkana, Kenya).

The species is characterized by a cranial capacity more significant than that of *Australopithecus africanus* and close to that of *H. habilis*. The massiveness and robustness of the craniofacial complex were more marked than in *H. habilis*. The size of the jugal teeth was reduced, but that of the incisors was close to that of *A. africanus* and *H. habilis*.

Generally speaking, *H. erectus* is defined by apomorphic (derivative) characteristics: a cranial capacity between 750 and 1,250 cm^3, a flattened receding forehead, a supraorbital torus, an angulation of the occipital, a strong thickening of the cranial walls, a marked sagittal tracheal spur, a developed occipital torus, a short and broad face and a weak chin. The plesiomorphic (archaic) characteristics were manifested by a flat head, a pentagonal cranial contour in posterior view with maximum width toward the base, a total and alveolar facial prognathism and a large set of teeth.

Between 600,000 and 300,000 years ago, the African and Euro-Asian continents experienced the radiation of new species of *H. erectus*. These evolved *Hominidae*, called archaic *Homo sapiens* or *Homo heidelbergensis*, were the subject of a major taxonomic debate, insofar as their characteristics were manifested by the persistence of plesiomorphies inherited from

Homo erectus sensu lato, and characteristics found in the form of synapomorphies in the lineages, both Neanderthal and Sapien. The hypodygma of *H. heidelbergensis* includes African specimens (Bodo and Omo in Ethiopia, Laetoli and Ndutu in Tanzania, Broken Hill, Kabwe in Zambia, Elandsfontein in South Africa and Salé in Morocco) and European forms (Swanscombe in England, Steinheim in Germany, Arago in France, Petralona in Greece and Sima de los Huesos in Spain; Hublin 2005). The new Moroccan remains discovered in 2017 in Djebel Irhoud could be part of this paradygma, since they are dated 300,000 years ago and are called archaic *H. sapiens* (Hublin *et al.* 2017).

The characteristics shared between these forms are represented essentially by a straightening of the frontal, an occipital tilt backward, the development of parietal bumps, a basic cranial flexion, etc.

Among the sites of Atapuerca in northern Spain that have yielded numerous human remains (Gran Dolina, Galeria and la Sima de los Huesos), Gran Dolina is about 800,000 years old and is the most significant in terms of lithic, faunistic and anthropological material. The archeological level TD6 has yielded several juvenile and adult human individuals, attributed by Eudald Carbonell and his collaborators to *Homo antecessor*. According to Arsuaga *et al.* (1999), the species evokes specimens mixing characteristics derived from *H. ergaster* and primitive characteristics of *H. sapiens*. The species is said to be at the origin of lineages leading to *H. sapiens* and *Homo neanderthalensis* (Bermudez de Castro *et al.* 1997; Arsuaga *et al.* 1999; Rosas *et al.* 1999). The skull of Ceprano, discovered in Italy, is as old as *H. antecessor*, pushing Manzi *et al.* (2001) to integrate it into the Spanish species, while Hublin (2005) for reasons of priority proposed the name *Homo mauritanicus* Arambourg instead of *H. antecessor*.

12.4. *Homo georgicus*

> *Homo georgicus* Gabunia *et al.*, 2002.
>
> Holotype: D2600.

The Dmanisi 5 skull (D4500, 2005) and its mandible (D2600, 2000) present massive and prominent superstructures, with an elongated and prognathic face similar to *H. erectus*.

It is in South Georgia that the oldest *H. ergaster* were found outside Africa, 1.7 My. Several skulls and mandibles belonging to youth and adults of both sexes demonstrate a mosaic of intermediate characteristics between *H. habilis* and

H. ergaster . The brain volume was close to that of *H. habilis* (between 546 and 730 cm^3). The mandibular body was constant from front to back and the symphysis was not quite receding backward, the size of the jugal teeth decreased from M1 to M3, the thickening of the dental enamel was present on the canines and the first premolar, lower limbs and vertebrae were adapted to walking and running, with a well-developed arch of the foot (Gabunia *et al.* 2000, 2002; Vekua *et al.* 2002; Lordkipanidze 2008).

12.5. *Homo neanderthalensis*

Spread across a hundred or so Eurasian, but essentially European sites, Neanderthals are today considered genetically well-defined, demonstrating a parental relationship with present-day humans to more than 2% of their genetic make-up. The oldest Neanderthals *sensu stricto* are probably those identified at Krapina, dated about 130,000 years ago. Other European sites, such as Chapelle-aux-Saints, Le Moustier, etc., are dated around 70,000 and 50,000 years ago, the most recent known at Arcy-sur-Cure, Saint-Césaire and Zafarraya, whose ages range from 33,000 to 36,000 years (Couture and Hublin 2005).

12.5.1. *Plesiomorphic and autapomorphic morphological features*

Paleontological work on Neanderthal cranial fossils is numerous and shows mostly the importance of plesiomorphic or archaic characteristics, the presence of autapomorphic characteristics which they hold from their origins, but few apomorphic or evolved characteristics, although many researchers claim the contrary (Heim 1976; Hublin 1978, 1988; Trinkaus 1983; Tillier 1988). The most remarkable plesiomorphic characteristics are: a significant volume of the face with a middle and upper level positioned high, absence of the canine fossa, significant facial and alveolar prognathism, hence the exaggeration of the craniofacial extension, voluminous and still very wide jugal teeth, shovel-shaped incisors, taurodontism (fusion of the roots), receding forehead, a significant supraorbital bulge developing a depression before the departure of the frontal scale, a low cranial vault and a low temporal scale developing in the same anterior-posterior extension of the vault (autapomorphic characteristics), occipital scale developing an occipital bun, as well as several other characteristics at the base of the skull (see below), the chin was weak, the craniofacial bones were massive and voluminous, reflecting strong musculature. Thus, the advanced characteristics can be summed up by the approximation of the cerebral volume.

12.5.2. *Non-Sapiens craniofacial dynamics and posture*

For a long time, researchers working on Neanderthals wanted to demonstrate a morphological and cognitive approximation with modern man, whereas all the characteristics of the skull and face, the occipito-cervical hinge, the curvature of the spine and its relationship with the position and adaptation of the pelvis and finally the posture adaptation of the knee in a dynamic of flexion demonstrated the opposite. Indeed, at the structural level, Neanderthal's dynamic did not reach a *Sapiens* dynamic. The facial field is very developed in front, and the craniospinal field even more so, retaining the squamous part of occipital bone raised backward. The absence of flexion of the occipital scale still keeps the slope of the clivus more upright, not favoring a mesial position and an elevation of the *foramen magnum* allowing the cervical spine to form a lordosis like all *H. sapiens* (Figure 12.2). In Neanderthals, the position of the *foramen magnum* led to very significant postural considerations. It must be noted that with such a position, the cervical spine could only be straight or kyphosis. If the cervical spine was in this position, the thoracolumbar spine assembly followed the same spinal posture. Thus, a straight or kyphosis posture unquestionably led to a lumbar lordosis insufficiency, because in this type of dynamic, the pelvis was in retroversion, thus reducing overlying functionalities such as lumbar lordosis insufficiency and putting in place underlying functionalities such as a semi-flexion leg posture. The recent reconstruction of the spine of the Kebara specimen in relation to the modern spine is part of this spinal logic of a kyphotic posture, reducing the physiological curvatures of modern man: cervical lordosis, thoracic kyphosis and lumbar lordosis. On the other hand, in *Sapiens*, in addition to the new characteristics that asserted themselves (forehead, chin, orthognathism and psalidodontia), the position of the *foramen magnum*, which is highly mesialized and slightly raised forward, responds to a new spinal adaptation, in this case a lordosis of the cervical spine (Figures 12.3 and 12.4).

12.5.3. *A permanent labidodental joint*

A recent work (Hadjouis 2015, 2020c) on dental joints of the genus *Homo* shows that, unlike modern man, Neanderthals had permanent labidodontia, in deciduous and definitive occlusion. The trajectories of craniofacial growth in the young Neanderthals (skull with an extended antero-posterior axis, absence or weakness of basi-occipital flexion, open sphenoidal angle, facial and alveolar-dental prognathism, etc.) were not modified in the adult stage. In the latter, the same rotations (spheno-occipital flexion, relation between synchondroses) and especially the same labidodental joint were found. Conversely, in modern humans, all children have a psalidodental joint, which appeared with *H. sapiens*, in conjunction with orthognatia, the formation of a straightened forehead and the development of a chin. This is no longer the case in young adults and adulthood, where both psalidodontia/labidodontia dental mechanisms have been regularly present in archeological populations from the Upper Paleolithic to the present day. Our orthodontists and dental surgeons are seeing in contemporary populations an increasing prevalence of bruxism and other oro-facial functions that put the dental joint into labidodontia.

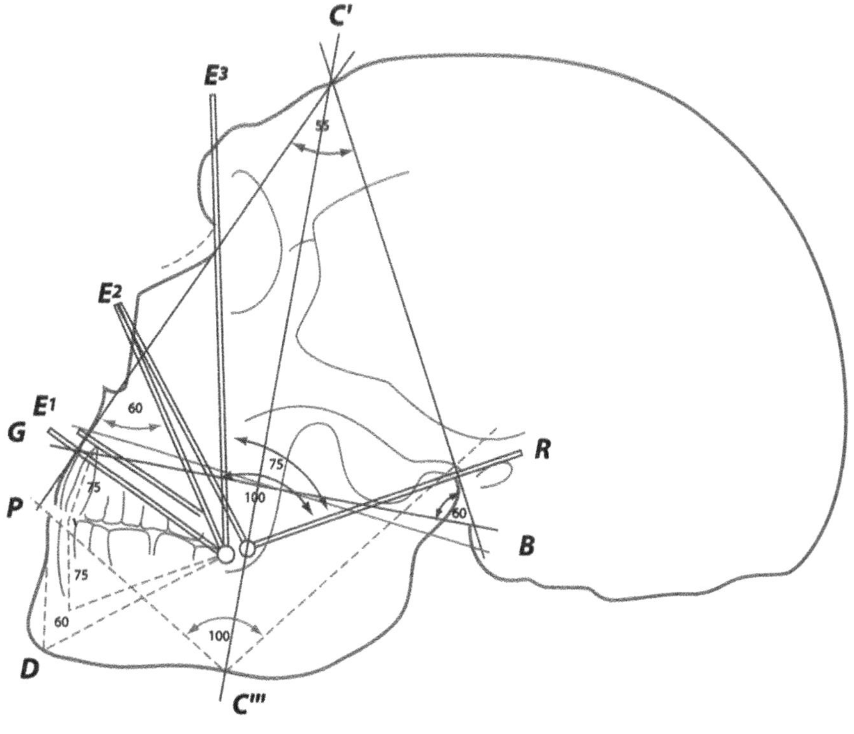

Figure 12.2. *Craniofacial and occlusal structural model of a Neanderthal (according to Leroi-Gourhan (1983) modified). For a color version of the figure, see www.iste.co.uk/hadjouis/skull.zip*

COMMENT ON FIGURE 12.2. – *The triangle closed at 55° translated during bipedal evolution, the development in height of the vault, the reduction of prognathism and the positive occipital tilt. In spite of the spheno-occipital flexion, the structural picture is still far from the Sapiens model, with an absence of development of the forehead and chin, and labidodental articulation.*

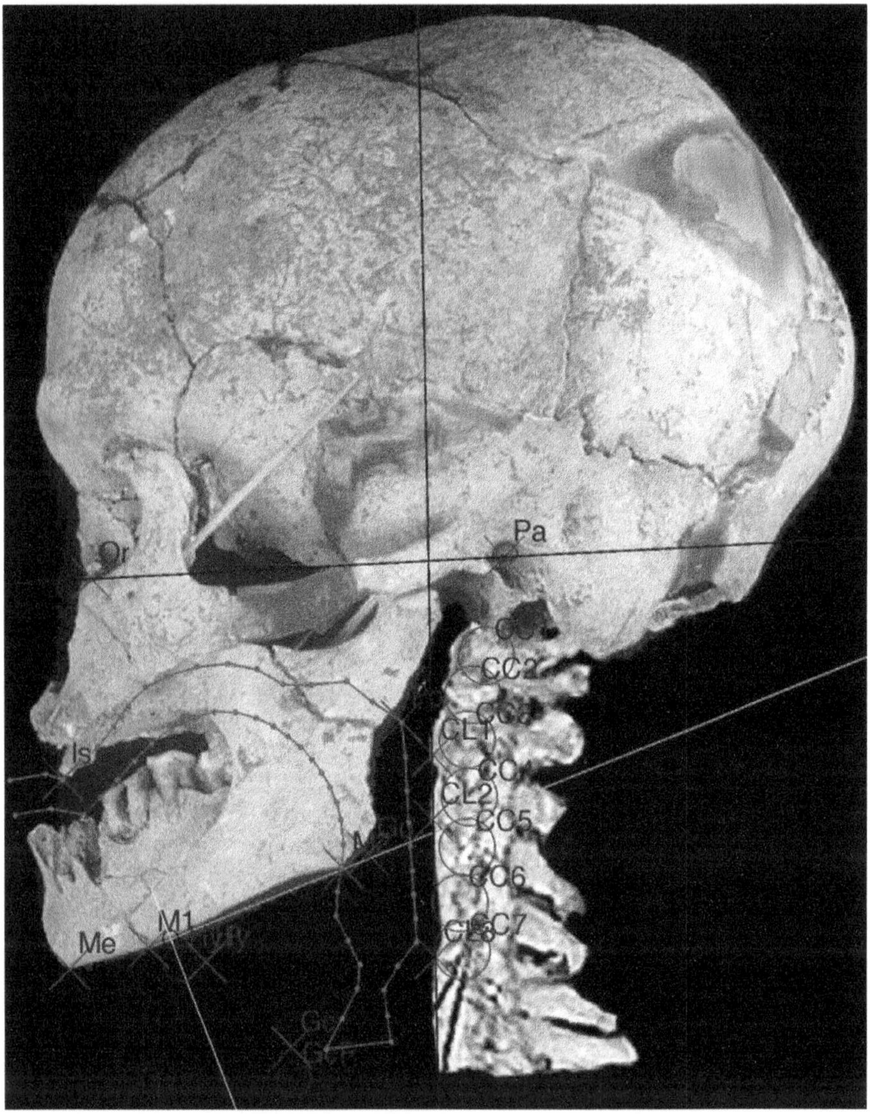

Figure 12.3. *Upper Paleolithic female skull from Taza (Algeria) and reconstitution of the cervical spine for vocal tract placement. Note the mesial position of the cervical spine (from Boé and Hadjouis (unpublished)). For a color version of the figure, see www.iste.co.uk/hadjouis/skull.zip*

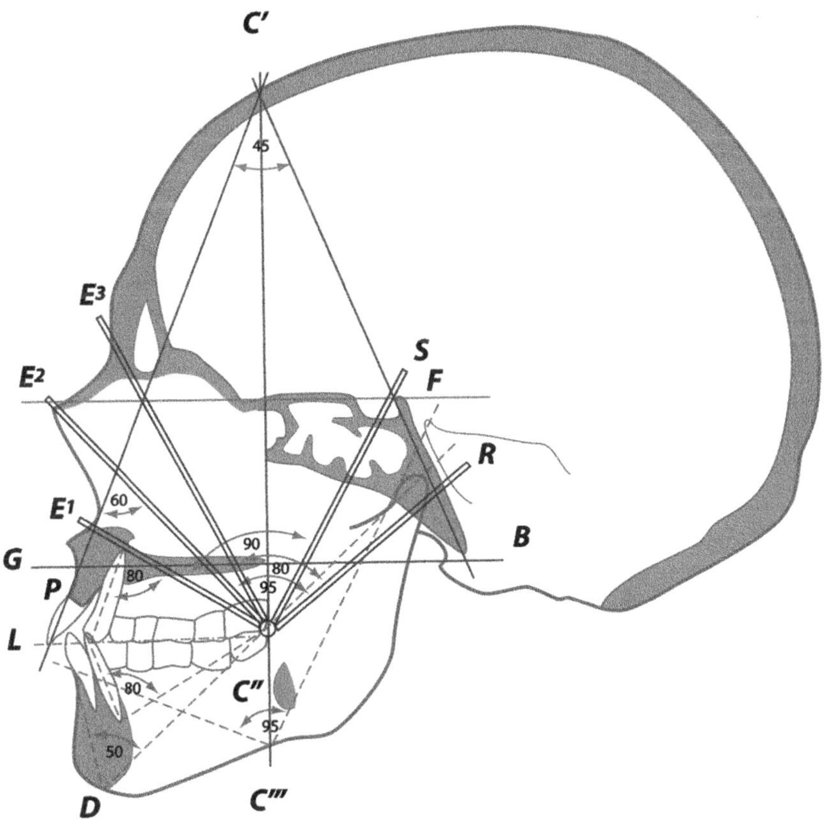

Figure 12.4. *Craniofacial and occlusal structural model of modern man (from Leroi-Gourhan (1983) modified). For a color version of the figure, see www.iste.co.uk/hadjouis/skull.zip*

COMMENT ON FIGURE 12.4 – *The triangle closed at 45° translated during the bipedal evolution of Sapiens, the development in height of the vault, the development of the forehead, an orthognathism, the formation of a chin, a psalidodental joint, an anterior occipital tilt and a strong spheno-occipital flexion.*

12.5.4. *The asymmetry of fossil pieces*

Great difficulties appear when taking measurements or describing skulls or mandibles deformed by a particular malformation during an individual's lifetime or by the weight of *in-situ* sedimentary deposits. The reassembly of brittle bones during the restoration of archeological remains is all the more problematic, since there is no evidence of the initial shape of the fossil in question. The virtual reconstitution of the *Sahelanthropus* skull from Chad is quite daring when one considers the large craniofacial deformation of the fossil, where the base of the skull appears entirely crooked, making the points far from their symmetrical axis.

In the same way, the reconstitution of the very damaged Neanderthal skull of the Chapelle-aux-Saints by Heim in the 1980s seemed indispensable both for the conservation of the piece itself and for the need to remedy the scientific errors it contained. The operation made it possible to re-establish a satisfactory symmetry (Heim 1998).

This is exactly where the mistake lies. What cranial characteristics in the reassembled fragments could show that the specimen was symmetrical? None, on the contrary, the fossil itself could have been affected by a malformation (craniostenosis, hemiplegia, etc.) whose symmetrical reassembly would hide the true cause. For a long time, with Jean Louis Heim, we raised this problem of deformation caused by a natural or pathological cause. Moreover, the base of the skull shows an important asymmetry of pseudo-plagiocephaly type, where all the characteristics of the left hemi-base are advanced with a rotation of the *foramen magnum*. The flattening of the left occipital bump is even visible from above.

Other fossil specimens have retained their deformation without any observations being made by the many paleontologists and paleonthropologists who have analyzed the piece. This is the case of the mandible of Tighennif III (Ternifine) *H. mauritanicus* discovered by Camille Arambourg in 1955 (Arambourg 1956).

This specimen was affected by a double left asymmetry (Figure 12.5). The first in facial view was a vertical asymmetry that had lowered the condylar process of the left mandible by about 1 cm relative to the right. This more important mechanical function on the left side raised the coronoid process by a posterior tilt, which appeared clearly higher, thus protruding beyond the condylar process. On the right side, the apophysis slightly exceeded the condylar process on the same side. Similarly, the sigmoid notch on the right was more stretched, more spread out, corresponding to the new mechanical functions of the second lateral asymmetry. The latter seen from above showed a significant lateral asymmetry, of the face torsion type, on the left. By tracing a tangent from the mandibular symphysis, the symphyseal axis made a significant rotation. Even in the absence of the *calvarium*, one can easily suggest that the middle level of the face was also affected by both types of asymmetry.

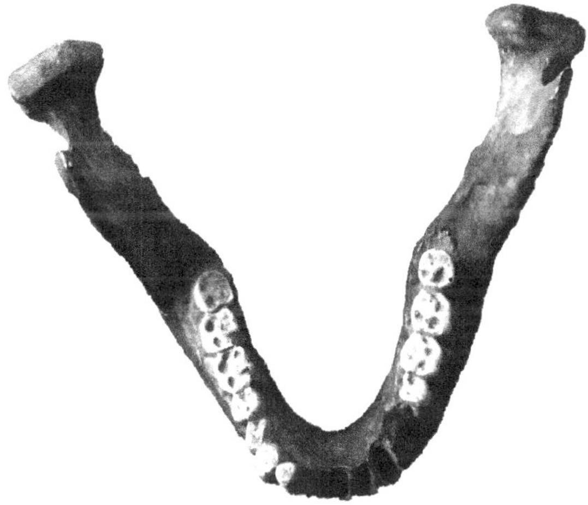

Figure 12.5. *Mandible of Homo atlanticus of Ternifine, seen from above. Note the strong asymmetry of the mandibular ramus. The vertical ramus and the right condylar process of the mandible have a translation followed by an antero-lateral condyle rotation. The right dental row was in a mesial position with respect to the antagonist row (© Hadjouis)*

12.6. *Homo sapiens*

The *Sapiens* species is considered to be the taxon that bears the greatest morphological and genetic resemblance to present-day humans. Compared to its predecessors, robustness has given way to gracefulness, sexual dimorphism is more marked and brain volume and cognitive abilities are significant. However, it is more with regard to craniofacial morphology that the differences are clearly more marked.

The build is new in terms of structural designs. Of all the *Hominidae*, it was the first to experience a reduction of the anterior-posterior craniofacial axis thanks to new mechanisms of trajectories and growth rhythms that have allowed the face to regress significantly (orthognathism), a development of the chin, a psalidodental joint, a heightened development of the forehead, an elevation of the vault, anterior tilting of the squamous part of the occipital bone, a mesialized *foramen magnum* while rising forward in response to cervical lordosis, as well as a large number of changes in the deep structures of the basic cranium.

Imagery of the base of the skull reveals adaptive and postural capacities that neither the "Archaic *Sapiens*", nor the Neanderthals, much less the Ante-Neanderthals could acquire, because the model of the structural mechanism "occipital tilt and articulated psalidodental joint" had not yet been realized. These two essential parameters are the result of kinetics of spheno-occipital, spheno-ethmoidal and ethmoido-frontal synchondroses. The new rotations resulting from trajectories during ontogenesis have allowed a large occipital flexion by a firming of the sphenoidal angle, a verticalization of the clivus, an advance and descent of the glenoid cavities, a craniofacial field, equivalent to the craniospinal field and a frontalization of the petrous pyramids.

Unfortunately, the evolution of *H. sapiens* did not bring only good adaptive things. While the labidodontia of the previous species of the genus *Homo* did not give rise to any malocclusion, the psalidodental joint, a new occlusal mechanism, caused class II and class III dysmorphoses due to slower growth rates (weak anterior rotation and occipital flexion, opening of the sphenoidal angle, increase in the craniospinal field in favor of the craniofacial field, etc.), responding with a retro-mandibulus with or without promaxillia) or at accelerated rates (large anterior rotation and occipital flexion, closure of the sphenoidal angle, reduced craniofacial field in favor of the craniospinal field, etc., responding with a protrusive mandible with or without retromaxillia).

In reality, the psalidodental joint came to represent for *H. sapiens*, the site of innumerable occlusal upheavals, representing both the effect and the cause. Nocturnal oral breathing, lingual play, swallowing, thumb sucking in babies, facial asymmetry, unilateral chewing, etc., are all factors of the oro-facial sphere, added to the infracranial factors whose postural imbalance is manifest in bipeds.

Describing these new craniofacial and occlusal structural kinetics of the human species, which the majority of researchers do not take into account, is ultimately due to the taxonomic misunderstanding of the status of the species. Indeed, in the absence of *H. sapiens'* diagnosis, only the polarity of morphological characteristics could function (apomorphy, synapomorphy and autapomorphy), whereas the innovative mechanisms of the *Sapiens* model and their consequences have been overlooked. The example of the new species of Djebel Irhoud (Morocco), dated 300,000 years ago, discovered in 2017 by Hublin and Ben Ncer is symptomatic. Several features of this specimen, such as a low vault and the extended shape of the neurocranium, induce a weak anterior rotation of the occipital and neck, which reflect the insufficient basal-occipital flexion. At the facial level, the prognathism is imposing and the dental articulation is still labidodental. The supraorbital ridge is prominent.

However, the oldest *H. sapiens* in Africa (195,000 years and 157,000 years) before the discovery of Moroccan specimens, notably the Ethiopian fossils of Herto man (*Homo idaltu*) (White *et al.* 2003), discovered in the Afar in 1997, and those of Kibish discovered in the Omo valley had modern characteristics, except for a robustness that is still present.

13

Migration and Paleogeographic Distribution of the *Homininae*

13.1. *Australopithecus* and *Homo habilis*: regional African migrations

In 1983, Yves Coppens proposed his theory of the African origin of man called *East Side Story*. It is based on a very simple idea: in the tectonically unstable regions of East Africa, the formation of the Great African Rift (Rift Valley) would have led to the creation of two environments with distinct climates and the establishment of two groups of *Hominidae*. The Great Apes would have been differentiated in the forests of the West, whereas in the East, the *Hominidae* would have evolved in a savannah environment. Although this theory no longer holds water, even acknowledged by its author, the African continent is still the birthplace of humanity.

The 1990s and 2000s saw the appearance on this continent of discoveries of great importance, raising the question of the origin of the lineage of *Hominidae*. Indeed, in the heart of the Sahelian country and far from the classic discoveries of East Africa, the new fossils that appeared in Chad were not only older but also located in regions that were previously unsuspected. The fossils of Abel found in Chad (*Australopithecus bahrelghazali*) and Toumai (*Sahelanthropus tchadensis*) are a source of hope for paleontological research, whose archeological exploration is increasingly concentrated in the West without abandoning the sites in the lake region (Ethiopia, Kenya and Tanzania) and those of Sterkfontein and Makapansgat in South Africa.

However, attention is now focused on new geographic regions such as the Maghreb. Initially reputed in Prehistory for having delivered the oldest *Homo ergaster* (*Homo mauritanicus*), found in the Acheulean site of Tighennif (Ternifine), then recently in 2017, by the discovery in Morocco (Djebel Irhoud) of the oldest *Homo sapiens*, the Maghreb could soon deliver their most ancient *Homo habilis* or equivalent *H. sapiens* in very promising sites in the Setif region (Ain Hanech, El Kherba and Ain Boucherit), where this last archeological site delivered dates around 2.4 My (Sahnouni *et al.* 2018).

Unlike modern man (*H. sapiens*), the only human representative on the Earth, the biological history of the *Hominidae*, bushy evolution and nonlinear, has gathered a large number of fossil species, some of which have coexisted together.

The evolutionary history of true bipeds began with the *Australopithecus* group of the family *Hominidae*, a link between the Primates of the Miocene (Tertiary stage) and the species of the genus *Homo* of the Pliocene (last stage of the Tertiary). The first species of this group to attempt an adventure outside the forest domain was *Ardipithecus ramidus* (4.4 My), a species whose remains were found in 1994 and again in 2001 by the American paleontologist Tim White and his team in the Awash Valley in Ethiopia. Other slightly more recent *Australopithecus* species ranging from 4.2 My to 3 My were found in Kenya, Tanzania and Chad. Among them were *Australopithecus anamensis*, *Australopithecus afarensis* and *A. bahrelghazali*. In South Africa, the species *Australopithecus africanus* is known from 3 My. From 2.5 My other species of the same genus are known such as *Australopithecus garhi* (Ethiopia), *Paranthropus aethiopicus* (East Africa), *Paranthropus boisei* (East Africa) and *Paranthropus robustus* (South Africa). These last two, dated around 1.8 My, are synchronous with the first species of the genus *Homo* and probably cohabited in the same regions.

In recent years, two important *Hominidae* discoveries have made the headlines in terms of the age of the species (6–7 My for the one and 6 My for the other) that place them well before the *Australopithecus* group and their environmental origin. The first, Toumai or *S. tchadensis*, is a skull found by French paleontologist Michel Brunet and his team in 2001 in the Djourab desert in Chad. Its skull, the smallest of all fossil *Hominidae* (360 cm^3), presents both archaic (long and narrow base, powerful nuchal musculature) and evolved characters (anterior position of the occipital hole, with probable bipedal locomotion, reduced and early canine). The second *Orrorin tugenensis* is represented by mandibular fragments and portions of the femur, first found by Kiptalam Cheboi, Brigitte Senut and Martin Pickford (Franco-Kenyan team) in 2000 in the Lukeino Formation in Kenya. The teeth of this *Hominidae* were smaller with a frugivorous or omnivorous diet and the lower limb (femur) curiously presenting a bipedal mobility close to humans and which distances it from *Australopithecus*. In any case, this characteristic shows an evolutionary scenario completely different from the previous versions, making this *Hominidae* an acceptable ancestor of the genus *Homo*.

The period between 2.5 and 1.8 My is complex for paleontologists, because not only was biological humanity making its entrance with the genus *Homo* in East and

South Africa (*H. habilis* and *Homo rudolfensis*), but we could also find *Australopithecus* which had not completed their evolution like *P. boisei* and a new species called *H. ergaster*, the oldest *Hominidae* of the human race recognized on a complete skeleton of a teenager found in Turkana.

The migrations concerning this pivotal period of the *Hominidae* did not go beyond the African continent and were only known in certain regions (East Africa, South Africa and the Sahel). The discovery of fossils in recent years in Chad shows that other African regions that have not yet been explored or have not yet yielded ancient human remains are on the list of sites to be explored, all the more so since their exit from Africa took place in North African migration corridors in East and perhaps West Africa.

13.2. *Homo ergaster* and *Homo erectus*: the first great African-Eurasian journey

All paleontologists believed that the exodus of *Hominidae* from Africa by *Homo erectus* occurred because of a greater brain capacity able to develop more efficient tools and able to conquer new geographical areas. We know that today this hypothesis no longer holds since the oldest *Hominidae* in Europe or Asia are as old as the last African *Homo* and that their morphology as well as their lithic technology and other hunting techniques had not evolved either. The example of the man from Dmanisi in Georgia is the most characteristic.

The first migration out of Africa was around 1.8 My after reaching North Africa. Algerian sites between 2.4 My and 1 My have revealed remains shaped by *H. habilis* or *H. ergaster* at Aïn Hanech and Aïn Boucherit in the region of Sétif and *H. ergaster* at Tighennif near Mascara in Algeria.

Until recently, the oldest inhabitants of Europe were dated less than 1 My. Classical sites such as those of Tautavel in France are dated at 450,000 years old. New excavations in Ceprano, Italy and Atapuerca in Spanish Catalonia have yielded human remains of *H. erectus* 800,000 years old and around 1 My at another named *Homo antecessor* in the north of Spain. Spanish researchers consider moreover that their fossils could represent a European lineage which would have gradually evolved toward Neanderthals.

Today, with the multiplication of archeological excavations, the discoveries are numerous and some of them greatly modify the recently established scenarios. The bone remains found in the Transcaucasus in Dmanisi (Georgia) are not only very old (1.8 My) but also represent several individuals, rarely found together elsewhere in the same site at that age. *Homo georgicus* thus represents the oldest European. This

H. ergaster with its still archaic morphology is astonishing because we should have in Europe individuals with a more modern anatomy. Some skulls resemble *H. habilis* (face projected forward, rounded contour of the back of the skull) with a cranial capacity of about 600 cm^3. The vertebrate fauna accompanying the human remains had African affinity (saber-toothed tiger, *Giraffidae* and giant ostrich) and suggests that the humans of Dmanisi followed the same migratory movements of African fauna from North-East Africa through the "Levantine corridor" to the Far East, since the Georgian site of *Homo* is the only one in Europe to have delivered African species.

Between 500,000 and 900,000 years ago, Europe had a population that was only slightly different from the *H. erectus* of Africa, but still had ancestors from that continent. These classical *H. erectus*, also called Anteneandertalians (or *Homo heidelbergensis* or *H. antecessor*), were known in Mauer and Steinheim (Germany), Montmaurin, Tautavel, in the Lazaret cave, La Chaise, Biache-Saint-Vaast (France), Ceprano, in the Prince's cave (Italy), Atapuerca, in the Sima de Los Huesos in Gran Dolina, Galeria (Spain), Petralona (Greece) and Azych (Azerbaijan).

In Asia, the hypotheses of the origin of Far Eastern settlements and their replacement are also ancient but much debated. In the Indonesian archipelago and particularly in Java, the first *H. erectus* of Modjokerto and Sangiran 31 (the Pithecanthropes of Java) are dated, respectively, at 1.8 and 1.66 My and support also an African origin. According to the replacement theory, the evolved *H. erectus* are represented on the island of Java at Ngawi, Ngandong and Sambungmacan.

In China, the first anthropological discoveries date back to the beginning of the last century with the knowledge of the Peking man whose remains (teeth) came from a limestone exploitation in the Zhoukoudian cave. The Sinanthropes (archaic *H. erectus*) are just as old as those of Java but the hypotheses diverge on the evolution of their Asian population. Rather, they are in favor of the continuity theory, which states that modern Asian populations are direct descendants of the earliest settlements *in situ* (Wei Dong 2008). The oldest remains found in China come from Yunnan Province at the Yuanmou site dated 1.7 My. However, the chronological position of the fossil remains did not seem to come from the Lower Pleistocene and would be of the order of 600,000 years. *Homo erectus* from the Lantian district and Hubei province in central China also provided human remains dated between 500,000 and 880,000 years ago. Other evolved *H. erectus* are between 300,000 and 120,000 years old and have been found in the Dali, Hexian and Maba sites.

With *H. erectus*, it can be said that with the exception of the Americas, all regions of the Old World of Africa and Eurasia were colonized. This was not the case for Neanderthals, discovered only in Europe and the Near East, apparently blocked by the ice caps of the first half of the Upper Pleistocene.

13.3. *Homo neanderthalensis*: a Eurasian traveler

While *H. ergaster* was typically African, Neanderthals were considered to be European and classical Near Eastern. The former experienced migrations within Africa, the latter, its population movements, were mostly localized in Europe around a little more than 100,000 years ago. Later, they were found in the Near East and Asia, mainly in Israel.

The ancestors of Neanderthals were European Anteneandertalians; however, paleoanthropologists diverge on the European lineage. Spanish paleoanthropologists, those with the largest number of bone remains in Europe, seem to be divided on the Neanderthal origin. Juan Luis Arsuaga of the University of Madrid considers that the *Hominidae* of la Sima de Los Huesos, related to the species *H. heidelbergensis*, are the ancestor of Neanderthals. For José Maria Bermudez de Castro it is *H. antecessor* in Gran Dolina that seems to be the right ancestral candidate. For the latter, this species is the common ancestor of all the Anteneandertalian and Neanderthal Europeans.

For Rigthmire of New York University, *H. heidelbergensis* is the species that gathers both European and African fossils and would be the common ancestor of Neanderthals and modern humans.

The body morphology of these *Hominidae* was robust and massive. In terms of the skull, the brain volume increased considerably with an average of 1,600 cm^3, the forehead receding with a strong bulge above the orbits and the body was stocky. They were first found in the eponymous site of Neanderthals (Nean-der valley near Dusseldorf), in Charente (site of La Quina near Angoulême), La Chapelle-aux-Saints, Mont Circé and Saccopastore in Italy, Spy in the province of Namur (Belgium) and Gibraltar (Great Britain).

Western Neanderthals pushed their eastern migration to Uzbekistan represented in the Teshik-Tash site and Shanidar cave in Iraqi Kurdistan. Halfway through this long eastward journey, large numbers of Neanderthals settled in the Middle East, especially in Israel. The Kebara cave, located in Mount Carmel and that of Amud near Lake Tiberias, has yielded several skeletons dated between 50,000 and 60,000 years ago.

13.4. *Homo sapiens*: the second great conquest voyage on all continents

The idea of a parallel evolution between modern man and Neanderthals is not new. Indeed, for a long time, European paleoanthropologists supported the hypothesis that Neanderthals represented the ancestor of *H. sapiens*; however, certain excavations in France and the Near East have showed rather a cohabitation between two species. The site of Saint-Césaire in Charente Maritime, excavated by François Lévèque, made a revolutionary discovery in 1979. A Neanderthal skeleton lay in archeological levels of the Upper Paleolithic: the Chatelperronian (36,000 years old).

The Franco-Israeli excavations that took place at the same period showed a probable encounter in the Near East between Neanderthals and modern humans. Some emigrated eastward from Europe, others, arriving from Africa, settled in Skhul (south of Haifa) and Qafzeh. When the latter arrived in the Near East around 100,000 years ago, Europe was still inhabited by Neanderthals. It would take 60,000 years before modern man invaded Europe during the Upper Paleolithic.

In Africa, between 100,000 and 200,000 years ago, the first archaic *H. sapiens* appeared with modern physical features (Omo and Herto valleys in Ethiopia, Laetoli in Tanzania, Klasies in South Africa and Taforalt in Morocco). The limbs as well as the waist became slender. In terms of the skull, the forehead was high, the prognathism disappeared little by little and the torus above the orbits was diminishing. It is these characteristics that we find 35,000 years later, more modern, in Europe and throughout the Old World. However, some of these characters seem to have evolved differently in the most archaic specimens of *H. sapiens* of Jebel Irhoud, dated 300,000 years ago.

These questions become more interesting when approached with a multidisciplinary approach. Indeed, paleoenvironmental and faunistic contributions show that modern man lived in Africa under a warm climate and hunted savannah fauna. When they arrived in the Near East, they found a somewhat similar climate and some of their game had an African affinity. The climate in Europe, 35,000 years old, was temperate and did not finally represent an ecological barrier for modern men coming from the East.

It was in France that the first classification was given to modern humans called Cro-Magnons. The Cro-Magnon shelter (Les Eyzies in Dordogne) found in 1868 by workers contained in Aurignacian levels, the skeletons of four adults and one child all bearing modern morphological characteristics (high skull, high forehead, vertical face, wide and short, absence of torus above the orbits, low orbits, developed chin, lengthening of forearms and legs, large waist, etc.). Other skeletons were also discovered in the Dordogne (Abri Pataud, Chancelade and La Ferrassie) and elsewhere (La Quina, Fontechevade and La Chaise). At the Italian border, Grimaldi men were found in several caves.

Genetic and paleontological work shows that the exchanges between Africa and Eurasia have been repeated several times. Mitochondrial DNA and Y-chromosome data suggest that between 150,000 and 80,000 years ago, African populations invaded the rest of the Old World according to the Out of Africa theory (Templeton 2002). It is these exchanges of African, Near Eastern, Eurasian, Australian and American populations that allowed the biological species *H. sapiens* to survive.

PART 3

The Skull of *Homo sapiens* in All its Diversity

14

The Craniofacial Puzzle in Motion

14.1. Normality and its boundaries with the abnormal and the pathological

The notion of normality lavished many times by Dr. Dastugue is not followed by paleontologists and paleoanthropologists. This is not done out of tactlessness or disdain, but rather because this notion is not taken into account in the academic teaching of these disciplines, to the point where the abnormal is sometimes considered pathological. If this last point seems somewhat exaggerated, the fact remains that the boundaries of the normal, the abnormal and the pathological, defended and explained by the said doctor, are simply ignored.

But what is normality? In human biology, this notion is variously appreciated according to the morphotypes resulting from human diversity; however, the vagueness persists because sometimes it is appreciated in relation to a statistical mean, with a variance and a standard deviation (frequency of individuals), sometimes it is functional translating a function without obstacle, without constraint and with a good muscular balance (Langlade 1981). This last author completes the examples of this normality by the ideal norm corresponding to a notion of anatomical order (esthetic ideal, purely theoretical because it is non-existent in nature), gnathological (alignment of the teeth, occlusion in centered relation, etc.) and periodontological (normal abrasion for self-balancing) (Langlade 1981). It is necessary to admit that in spite of the examples taken in such or such other discipline the notion of normality remains fragile taking into account the presence within Mammals of strong polymorphism and dimorphism because each individual remains a biological entity of its own.

Mastering at least the norms of normality, the identification of an abnormal character (not usual in the normality of an organ, occlusal surface, a given joint) but not pathological (infectious, degenerative, metabolic, neurological, etc.) could help in the advancement of the bone piece to be determined. On the other hand, avoiding the classic adage of former physicians "everything that is not normal is pathological" will do a lot for retrospective diagnosis. This expression, often cited by Dr. Dastugue, shows the foot orthopedist who he was before he became a paleopathologist, the rigor of his discourse in the face of such statements.

Thus, if the abnormal is defined as the exception to the norm, causing no damage to the wearer either morphologically or functionally, "the pathological is what affects, what one sometimes feels painfully, and which ultimately causes damage" (Dastugue and Gervais 1992, author's translation). In other words, pathological bone affected by lesions of an osteoplastic (new bone formation) or osteolythic (bone destruction) nature will not go unnoticed, even less so before the trained eye of the expert in paleoanthropology or paleopathology.

14.2. The importance of interpreting or reinterpreting (Le Double 1903, 1906)

A renowned anatomist, but quickly forgotten because he died young (1848–1913), Anatole-Félix Le Double published four volumes devoted to the muscles and bones of the skull, face and spine. Doctor Pierre-Léon Thillaud, who republished his work *Traité des variations anatomiques ... et de leur signification au point de vue de l'an-thropologie zoologique* at the Bibliothèque interuniversitaire de Médecine in 2004, will said of him:

> This was his "Great Work" and of all his innumerable and fascinating productions, never before had descriptive anatomy pushed so far the necessary conjugation of his essential talents: the sharpness of his gaze and the abnegation of his observation, applied to morphological detail. In truth, no anatomist will venture to follow up on this solitary and monumental entity. (Thillaud 2004, author's translation)

Fortunately, there is renewed interest in the works of this anatomist, so particular that the anatomists of his time already referred to him as "the man of variations". Paleopathologists and osteopaths were the first to be interested in the wealth of information provided by the "Treatises of Variations".

Process, crest, furrow, roll, bulge, tuber, dimple, facet, canal, concha, cone, hook, notch, groove, etc., are all terms used in the nomenclature of anatomical variations of the skull and face of humans, as well as in that of the comparative anatomy of Primates and some other Mammals.

It goes without saying, as pointed out above, that the slightest anatomical character by its osteoplastic or osteolitic topography is described precisely, in the light of its normality or abnormality.

Several of these characters also represent what paleoanthropologists call discrete characters such as suture bones, supernumerary bones, the presence or absence of orifices and grooves, vascular or nervous (Crubézy and Sellier 1990).

14.3. Craniofacial structural mechanics and dynamics

In order to make the mechanical and dynamic interpretation of the different pieces of the craniofacial puzzle more legible during ontogenesis, the biometric analysis of the skull, face and teeth has been completed by the implementation of a new metric and descriptive method digitized through medical imaging (Hadjouis 2001). The latter, first in two-dimensional profile and then in three dimensional, illustrates the relationships of deep structures on a ground in harmony or disharmony. Several phases are thus described, from dental analysis (anomalies and oral pathologies) to the determination of the structural picture (dental and skeletal dysmorphoses) through biometric structural location by radiographic analysis. Knowing that teeth come through only after craniofacial formation, their study (developmental abnormalities, dental movements, etc.) in relation or not to maxillo-mandibular imbalances is essential in both temporary and permanent dentition.

A large part of this method is devoted to the relationship between dental movements and rotations and occlusion. These are the movements of gression, version and rotation, analyzed clinically at first and then during teleradiography in relation to dysmorphoses. Certain optional sections exist and touch on certain forms of ritual practices found in prehistoric populations (dental avulsions, mutilations, etc.). Similarly, the practices of dental parafunction, i.e. the use of teeth as a tool (tanning of skins by the anterior teeth, maintenance of spinning, etc.), give the occlusal surfaces a profile that distances the usual dentine articulation.

But it is above all the sections concerning the biodynamics of the skull and face and its relationship with the occlusion that represent the essence of the method. Each skeletal piece of the craniofacial puzzle is analyzed in relation to the harmonious or disharmonious development of the cranial sutures, on the one hand, and the maxillo-mandibular balance or imbalance, on the other hand (Figures 14.2–14.4). Thus, such a biodynamic study is based on two indispensable steps: a macroscopic analysis of the rotation of each bone and an analysis of the metric and angular ratios of the flexion and extension dynamics by teleradiography (Figure 14.1).

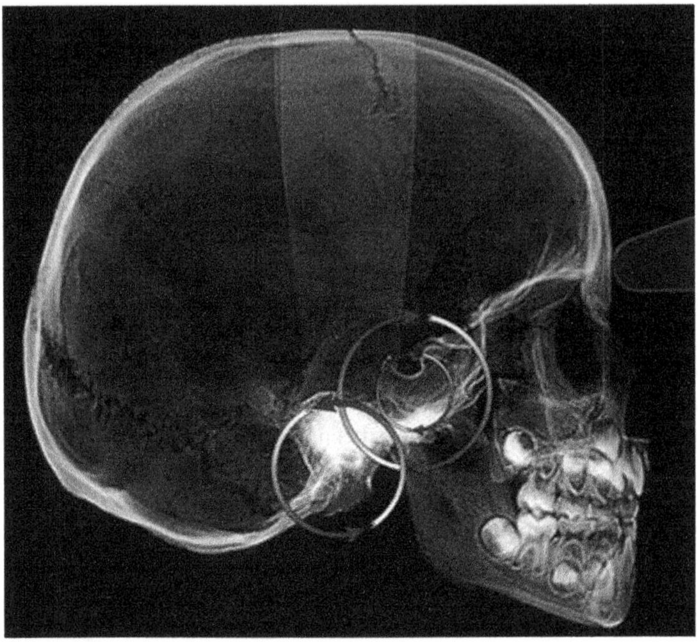

Figure 14.1. *Dynamic pattern of spheno-occipital kinetics during growth. The basi-sphenoid rotates clockwise, while the basi-occipital rotates counterclockwise, teleradiographic profile photograph of a child from Chevilly-Larue necropolis (Val-de-Marne) (© Hadjouis, Katz). For a color version of the figure, see www.iste.co.uk/hadjouis/skull.zip*

14.3.1. *Biodynamics of vault bones*

As we have shown in the previous pages, the development of the vault, whether normal, abnormal or pathological, depends on the sutural formation of the skull, its itinerary, abnormalities or early closure. In fact, isolated, mixed or secondary to other particular pathologies craniostenoses have an important part in the development of craniofacial disharmony. The asymmetries caused by craniostenoses are not the only ones; many other lesions affect craniofacial balance and symmetry. Thus, in neurocranial abnormalities, hydrocephaly, such as microcephaly, found in some populations also alter the dynamics of the skull.

In some archeological populations, there were also two types of cranial deformations, one artificial, the other voluntary. The first one was generally caused by cultural habits such as swaddling or bandaging of infants, their mode of lying causing significant flattening of the outer table (frontal, fronto-occipital, etc.). The second, whose main type was the Toulouse one, its fronto-occipital flattenings and other depressions of the vault were much more impressive.

14.3.2. *Biodynamics of the temporal bone*

The temporal bone is one of the most important bony parts of the base of the skull for the understanding of cranial flexion-extension dynamics. Indeed, its internal or external rotation, the frontalization of the petrous pyramids and its relations with the other craniofacial elements make it possible to grasp the significance of structural relationships.

This part concerns the bilateral movements of the temporal scale in internal or external rotation or on the contrary their asynchronism, which is most often a sign of asymmetry. The description of the glenoid cavities (depth of the cavity), their orientation and the anomaly or pathology of the temporomandibular joint (TMJ) (modified joints, exostoses and eburnation) are the most significant criteria of a basic cranial asymmetry. The shape and position of the mastoid apophyses are also signs that will be added to the final diagnosis.

The frontalization of the petrous pyramids in external rotation will favor a large angular opening of the flexion type, whereas the movements blocked in internal rotation (small angular opening of the rocks, plating of the temporal scales) will have tendencies of the extension type. It happens that the dynamics of one of the two regions may be blocked, or in reverse rotation.

14.3.3. *Biodynamics of the occipital bone*

The movements of the squamous part of the occipital bone, the morphology and position of the occipital condyles, that of the occipital *foramen* and especially the state of the spheno-occipital synchondrosis (SOS) are all significant bony parts that record the slightest anomalies that appear during development. In this basic cranial dynamic, the spheno-occipital synchondrosis appears as the keystone of the understanding of the mechanisms between the facial skull and the spinal skull. The reverse rotational movements of the body of the sphenoid and occipital which are articulated at the SOS level represent two entities, one of which is basi-occipital and basi-sphenoidal, the other is sphenoidal and ethmoidal. In this articular context, the occipito-atloid hinge, i.e. the craniospinal relationship, is crucial.

14.3.4. *Biodynamics of the sphenoidal bone*

Like the occipital bone and the tilting movement of its scale, the sphenoid also takes on a special character due to the flexion of the clivus. The measurement of the sphenoidal angle, put in parallel with the other measurements, allows even before the occlusal analysis to suggest the structural type and adaptation of its bone structures. The position and morphology of the clinoid processes will be essential to help to grasp the meaning not only of the phenomena of descent or ascent around the sella turcica, but also at the level of the spheno-ethmoid-frontal planum.

14.3.5. *Biodynamics of the maxillary bone*

Björk and Skieller (1972) have shown that jaw growth is sagittally stable. For these authors, maxillary protrusion in the face depends not only on the amount of growth in length, but could also represent an overall effect of maxillary displacement. While vertically, Björk suggests that maxillary growth occurs at the frontal and zygomatic sutures in association with tooth eruption by inferior apposition of the alveolar processes.

The naso-maxillary complex moves transversally thanks to the internasal and intermaxillary medial sutural movement. Ricketts and Langlade suggest the existence of two polar centers, left and right, located at the level of the large round holes. This hypothesis would provide an explanation for the phenomena of symmetry and asymmetry (Langlade 1981).

The maxilla represents the middle level of the face. During life, if a facial imbalance due to malocclusion or any other phenomenon has set in during the crossing of the jaw, the signs will be more visible on the maxilla, as on the other levels of the face, because the oro-facial functions, in addition to the facial expression, are more expressive there than on any other bony piece that has developed the same disturbance.

These factors are related to the establishment of a good facial adaptation as well as a harmony of the face or on the contrary its disharmony. The topics in question concern the internal or external rotation of the maxilla, the asynchronism of the hemi-maxillae, the position of the maxilla and its relationship with the TMJ and the zygomatic arch. Other dynamic characteristics of the jaw bones are also significant, such as the position of the pterion, the position of the anterior nasal spine and the shape of the palate. The predisposition of the maxilla to alveolar protrusion or retroalveolus is discussed in the context of this form, as it will also be reviewed in detail during the analysis of maxillo-mandibular dysmorphoses. Furthermore, one must be careful about reversing growth trajectories in the same way as a discordant dysmorphosis that destabilizes the maxillo-mandibular complex, but in opposite directions (Figure 14.3).

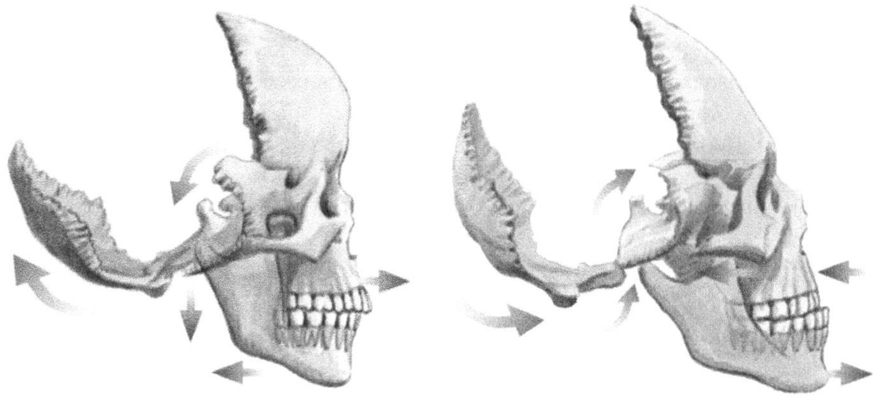

Figure 14.2. *Growth trajectories of the craniofacial puzzle during the development of a class II dysmorphosis (retromandibulia with or without promaxilla) and class III (protrusive mandible with or without retromaxillia) (© Tosello, Hadjouis)*

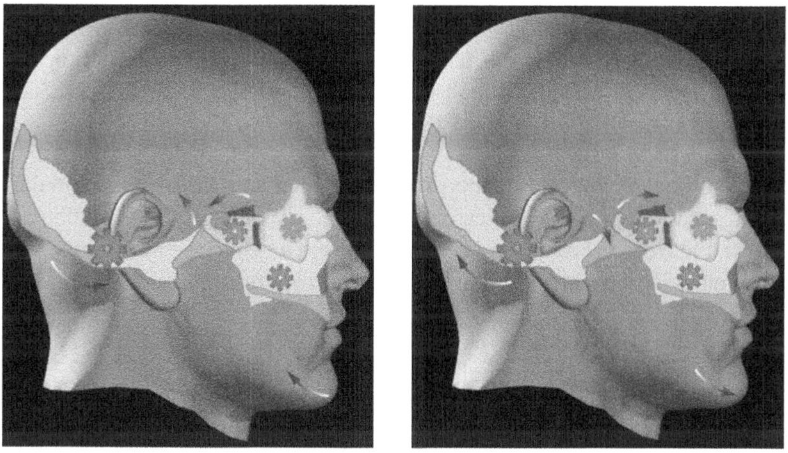

Figure 14.3. *The structural kinetics of discordant dysmorphoses, destabilizing the maxillo-mandibular complex, but in opposite directions (© Huard, Hadjouis). For a color version of the figure, see www.iste.co.uk/hadjouis/skull.zip*

The other sections concerning the face are devoted to disorders of the other levels in relation to many neurological and traumatic causes, etc.

14.3.6. *Biodynamics of the mandibular bone*

The lower level of the face is represented by the only mobile element of the craniofacial block which is the mandible. The mandible consists of the mandibular body and two rami rising on each side. The eminence of the chin at the front and the mandibular angle at the back indicate the antero-posterior limits of the body, also called the horizontal ramus. The rising ramus is composed of two processes, one in front is the coronoid process, the other in back is the condylar process which supports the TMJ. Mandibular growth occurs through condylar process displacement in the posterior and lateral direction. At the vertical and sagittal level, mandibular development occurs through the phenomena of apposition and resorption.

During growth, the mandibular angle is variable. From around 150° in newborns, it increases to about 120°–130° in adults. The value increases again in the elderly to reach 140°. This variation of the mandibular angle depends on the existence of the alveolar edge at the time of losing teeth in children and its loss in the elderly, but opinions are divided. Enlow (1968, 1973) believes that there is resorption of the anterior ramus to make room for the eruption of the second and then third molars as they erupt. For Moss (1969, 1974), the condylar process is adaptive to the environment of functional demand (chewing, swallowing, breathing and phonation). Ricketts (1977), while agreeing with Moss' ideas, proposes his theory of arch mandibular growth, emphasizing the apposition that occurs along the mandibular arch. In spite of this variation, the angle of the mandibular arch remains constant between 6 years and adulthood, as Langlade has shown. The posterior condylar process' increase is about 2.5 mm/year, while the transverse increase is about 2 mm/year (Ricketts 1977; Langlade 1981).

The mandibular proportions are provided by the angle of the mandibular arch, the mandibular length and the intercondylar process' distance. The mandibular angle remains stable from the age of 6 years in both sexes, while the mandibular length is shorter in class II (retrognathia), longer in class III (prognathia).

While increasing its value by about 2 mm/year, the intercondylar process' distance remains within a normal growth distribution. However, the distance is shorter in class II, greater in class III.

Whether vertical or sagittal, the TMJ can be the cause or effect of facial asymmetries (torsion, lateral flexion, etc.) (Figure 14.4). Indeed, hypo or unilateral congenital hypocondylosis can cause significant vertical asymmetry of the face; conversely, hemiplegia or any other pathological or traumatic form will modify the condylar process' relationships as well as occlusion. It is by means of the condylar position, and more particularly the angle of the condylar axis to the right and to the left with the horizontal, that the angular value relationships are analyzed. The axis must be parallel to the auditory plane; any rotation, translation or changes of an infectious, inflammatory or traumatic nature suggest an ethiopathogenic diagnosis of asymmetry in the vertical, sagittal or both.

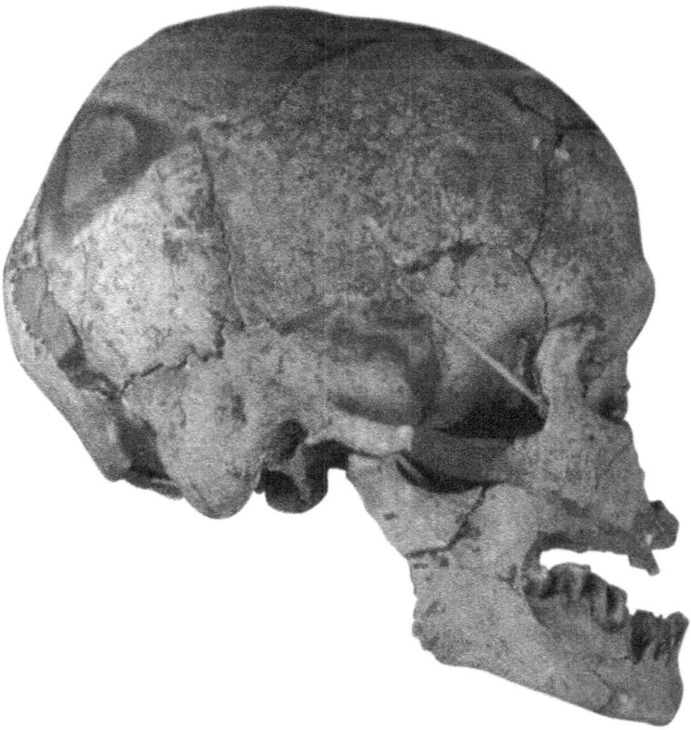

Figure 14.4. *Right lateral view of the skull of a Homo sapiens woman from the Upper Paleolithic of Taza (Algeria) showing a mandibular dislocation (© Hadjouis)*

Biometric analysis for a paleoanthropological study seems insufficient today without the contribution of a structural analysis to explain the functioning of dynamic mechanisms such as the occipital tilt and the processes that flow from it at the occlusal level. Clearly, craniofacial structure is different at each stage of ontogenesis, to the extent that what results from a simple dynamic movement or a change in orientation of a piece of bone could be considered as morphological variability. Examples are common, such as the temporal scale which changes its morphological criteria and metric values simply because of too much basic cranial flexion or, on the contrary, too little occipital tilt. These same values will change again under the influence of temporal asynchronism.

15

The Basics of Structural Analysis

15.1. Analysis tools using imaging

The structural and biodynamic analysis of the craniofacial puzzle that we practiced on our current and fossil collections was based on several protocols put in place by Delattre and Fenard, Gudin, Delaire, then Deshayes. In the context of his experimental work on the occlusal growth of young children, Deshayes revised all the methods and in particular that of Delaire (1978) in order to derive the essential parameters for the analysis of occlusion in relation to the skull and face and their ontogenic changes. It is finally this last method that is applied (Deshayes 2001) as a complement to the database we have created on maxillo-mandibular dysmorphoses in relation to the new dynamics induced by craniofacial malformations and their repercussions on the spine (Hadjouis 1999, 2000b, 2001, 2008b).

The main points of the structural analysis are presented in the following manner: AR: anterior rotation, ER: external rotation, MP: maxillary pillar, NA: nasion, NP: naso-palatine, Pcl: posterior clinoid, SPT: superior pituitary tubercle, SPT: superior pterygoid, IPT: inferior pterygoid, SOS: spheno-occipital synchondrosis, BA: basion, Op: opisthion, Ol: low opisthion and TC: temporal condyle.

To these points are added the calculation of angles and ratios. In this structural approach, the skeletal profile is no longer compared to an ideal craniofacial equilibrium line, but to predictive growth models (Deshayes 1986). Here, it is the kinetics of sphenoethmoidal and/or occipital rotation that remains crucial for any interpretation (Deshayes 1997, 2001). The latter proposes four planes of organization, inscribed within a cube delimiting the extreme points of the *calvaria* (former straight lines C1, C3, C4 and CF4 by Delaire):

– the sphenoidal-ethmoidal-frontal plane A follows the sphenoidal and ethmoidal planum;

– the spheno-occipital plane B or endocranial slope line of the occipital clivus and the basi-sphenoid;

– the occipito-pterygoid C plane is measured in algebraic value with respect to plane A;

– plane D is the craniofacial baseline.

All cranial collections from Europe and Africa that have passed through our hands have followed this protocol. Admittedly, this protocol is only effective in the presence of a complete or sub-complete craniofacial complex, but it is also effective on *calvarium* without mandibles. As a general rule, flexion of the base of the skull is followed by an elongation of the craniospinal field and a shortening of the craniofacial field, in which case the sphenoidal angle is closed, giving the clivus greater verticality. Conversely, an extending skull is followed by an increase in the craniofacial field and a reduction in the craniospinal field with an open sphenoidal angle making the clivus less vertical.

All fossils of Anatomically Modern Humans fall into the first category and inscribe their variability within this plane, whereas Neanderthal and Anterior Neanderthal fossil species fall into the second category. The sphenoidal and occipital structural kinetics are then evaluated according to significant cross parameters of dysmorphosis. In class II (skulls in extension) as in class III (skulls in flexion), four groups with two crossed spheno-occipital kinetics emerge.

While the dental articulation of pre-*sapiens* species remains exclusively locked in labidodontics (end-to-end incisors) in adults as in juvenile individuals, it is not the same in *Homo sapiens* where all immature individuals since the Upper Paleolithic have developed according to an exclusively psalidontal articulation, whereas the psalidodontia known with this species will deploy, according to the masticatory and para-masticatory actions, an abrasion of the occlusal surfaces making the incisivo-canine blocks in a labidodontal situation. This function was also present throughout history, aggravated today by the problems linked to bruxism.

Medical imaging technique such as teleradiography, although it is largely obsolete for some users because it only uses two dimensional, is still an indispensable tool for structural profile diagnosis, still supported today by Dr. Philippe Katz, one of the leading experts in the field. The new tomodensitometric parameters are also an essential complement to three-dimensional analysis.

Structural exploration using the teleradiographic process is very old since the first work was presented in 1922 by Carrea, followed by that of dentists specialized in dentofacial orthopedics (DFO) in 1936 and 1942. The most complete works concerning craniofacial angular variations in profile are first those of Gudin (1952), which are still relevant today, and then those of Leroi-Gourhan (1964), whose structural analysis concerned all vertebrates. However, a constructive interpretation of the historical reflection of craniofacial mechanics since the 20th century has allowed us to establish the fundamentals of structural dynamics of the craniofacial puzzle.

15.2. Maxillo-mandibular dysmorphoses

As a general rule, the distocclusion (or class II), the most important in contemporary populations, defines a distal position of the lower canine and M1. Characteristic features of this class include an exaggerated incisal overjet (division 1), where the upper incisors are oriented mesially and not vertically, or an incisal overlap (division 2 or overbit), where the maxillary incisors overlap the mandibular antagonists.

In mesiocclusion (or class III), rare functionalities in our current populations, with the exception of congenital, hereditary or syndrome-related dysmorphoses, it defines a mesial position of the canine and mandibular M1. The dento-mandibular projection in some individuals is such that it requires maxillofacial surgery. Similarly, the facial musculature (temporal, masseter, internal pterygoid and external pterygoid) is subject to forces and responding to the type of masticatory function. Here, the occlusal load is much greater on the incisors, and this labidontal activity involves the forces of the masseter and the internal pterygoid to the detriment of a temporal force. In contemporary mastication, class I or II incisal overlap involves the temporal muscle rather than the masseter.

The diagnosis and description of malocclusions followed by some orthodontists is based on facial typology. These authors, including Angle (1899), Sassouni (1959), Ricketts (1960) and Langlade (1981), propose a classification of facial types in relation to malocclusions. The terminology followed by these authors adopts the term "class" exclusively to dental malocclusion, while the term "type" is reserved for the typological facial or skeletal description. Thus, the terminology "normal skeletal" or "skeletal class I type" represents a normal facial type with vertical and sagittal proportions of the concordant face (Langlade 1981). Within these classes, large dysmorphic variations will occur as a result of facial disturbances (or anomalies) in the vertical and sagittal directions. Thus, in class I syndrome, the face is stable in the antero-posterior direction; however, the height is variable and can be revealed:

– in normal position vertically;

– of deep bite skeletal tendency (decreased facial height);

– skeletal open bite tendency (excessive facial height).

The malocclusion may present spaces that will reflect an alveolar protrusion or microdontia, or overlaps resulting from a retroalveolus or macrodontia.

In class II syndrome, the face is characterized by an antero-posterior maxillo-mandibular shift:

– of deep bite skeletal tendency;

– in normal position vertically;

– of skeletal open bite tendency;

– with a malocclusion with overlaps or gaps in one or both brow bones.

There are three situations that can occur at the skeletal bases:

– a mandible in normal situation and maxilla in front;

– a maxilla in front and mandible behind;

– a maxilla in normal situation and mandible behind.

In class III syndrome, the face presents a maxillo-mandibular offset:

– of deep bite skeletal tendency;

– in normal position vertically;

– of skeletal open bite tendency;

– with an overlapping malocclusion in one or both brow bones.

Three situations can also occur at the skeletal bases:

– one mandible in normal situation and maxilla behind;

– a maxilla behind and mandible in front;

– a maxilla in normal situation and mandible in front.

15.2.1. *Angle's classification*

Angle was the first to establish a typology based on dental occlusion patterns, in profile view. In class I-normoclusion, the cardinal sign is the relationship between the tip of the mesio-vestibular tuberosity of the upper first molar which is articulated in the bottom of the vestibular and central sulcus of the lower first molar (Skinazi 1995). Class I-malocclusion occurs when there are some irregularities due to overlaps or abnormal spaces. The occlusion relationship between the canines is such that the upper canine is always positioned behind the lower canine. The mesio-lingual side of the upper canine must be in contact with the disto-vestibular side of the lower canine. At the level of the incisors, the overjet between the upper incisal edge and the vestibular surface of the lower incisors should be more than 2 mm.

In class II, there is a half-cuspid forward shift of the upper molar in front of the lower molar. With this type of occlusion, the upper canine will be in front of the lower one with an upper disto-lingual side in contact with the lower mesio-vestibular side. Sub-classes also exist within class II (class II1 and class II2).

Class III is manifested by an inverse ratio of the first molars with, in particular, a half-cuspid backward shift of the upper molar with respect to its lower counterpart. The upper canine will thus be found behind the lower canine.

In addition, we should not neglect certain changes in the dental series modifying the occlusion, such as agenesis, early tooth loss or similar phenomena that position the active teeth in situations of gression (translation of the tooth) or version (inclination of the major axis of the tooth).

15.3. History of structural mechanics: from geometry to imagery

15.3.1. *The initiators*

The craniofacial structure of vertebrates has imposed on zoologists, paleontologists, anatomists and therapists, models of geometric construction based on the facial typologies of contemporary humanity and a nomenclature of the morphotypes of fossil species. These models make it possible to use drawings of the outlines of the face and the neurocranium, in order to locate the mechanical equilibrium between the base of the skull and the face by calculating a certain number of straight lines and angles, as well as to resort to in-depth analytical reflection between the base of the skull, its flexion during the stages of development and evolution, and the mesialization of the occipital hole during the straightening of the spine.

The major work on craniofacial structural mechanics is that of Leroi-Gourhan, carried out in 1955 on the whole of the Vertebrates, since the Placodermi fish of the primary era in a liquid medium, until Man passing by the Birds, Reptiles and Mammals. Although the geometrical drawing (the pantograph) is inspired by the work of Gudin in 1947 and 1952, the fact remains that the Sciences thesis, defended under the supervision of paleontologists and zoologists Piveteau, Grassé and Devillers, sheds a masterly light on the skull-face relationships. A considerable amount on the skull was born, under the pen of a prehistorian and paleontologist.

For this author, the mechanical representation of the cranial framework requires three devices: the first concerns suspension, the second traction and the third support. These three devices will be schematized by drawings and fixed points. The three construction drawings will be constant for all vertebrates: the basion (B), the external inion (IE) and the prosthion (P). The triangle of the cranial structure is materialized by a perpendicular line that divides the two basion-prosthion points into two equal parts: one dental, the other cerebral. It thus appears that the geometric center of the cranial construction corresponds to the last tubercle of the last molar (Leroi-Gourhan 1964, 1983). From the oldest vertebrate (Fish) to modern man, the triangle will tend to close as the orthognathic face in man takes shape, in association with a positive occipital tilt. In other words, the geometrical construction (at its upper compass) changed from an angle of 120° to 45°. This fundamental law in fact reflects constant mechanics that places the skull-face relationship midway between the anterior point of the occipital hole (B) and the anterior point of the dental alveoli (P). Any decrease in the length of the basilar back skull leads to an equivalent decrease in the dental fore-skull (Leroi-Gourhan 1983). This mechanical consequence, which, at first glance, seems simple, added the spinal framework, especially to its cervical segment during the transition to bipedalism, as well as the relationship between the vestibular and the jugal teeth.

While the geometrical layout put in place by Leroi-Gourhan relentlessly demonstrates the craniofacial-dento-spinal relationships, the same is true of the main previous works, notably those of Dr. Gudin. Indeed, this orthodontist who defended his two theses in 1951 and 1952, one, a dissertation from the *Académie nationale de médecine*, "*Contribution à l'étude du prognathisme ethnique par la téléradiographie*" (Contribution to the study of ethnic prognathism by teleradiography), the other, a Sciences thesis under the supervision of Professor Piveteau: "*Étude du crâne de profil : Anthropomorphes et Hominiens*" (Study of the skull in profile: Anthropomorphs and Hominians), concentrates his craniofacial structural study on the pantograph, invented by him. This cephalic pantograph represents a geometrical figure, schematizing the facial structures, delimited on an X-ray photograph by anatomical benchmarks. The basilar construction is connected to the vertebral structures by the angle formed by the spheno-occipital plane of the clivus with the foraminic plane BO (basion-opisthion). Thus, by exploring the angles of the basion, he found that progressive flexion of the base of the skull causes the expanding encephalon to unlock the fronto-parietal bregmatic fontanelle (Gudin 1952). In the same way, this geometrical construction makes it possible to reproduce and highlight the dynamics of the skeletal parts in correlation with the variations of the spine and vertical straightening (Gudin 2002).

All in all, it appears to Gudin that the construction of the facial mass is related to a certain number of dynamic factors linked to the base of the skull, among them: a prolonged spheno-occipital plane (clivian plane) representing a cranial base in contact with which the regression of the face took place when there was a decrease in ethnic prognathism. The movements of the sphenoethmoidal plane brought the facial line into its dynamicity. The closing of the sphenoidal angle, as all orthodontists see, was completed by the tilting movement of this profile line. It is the line of the clivus which, starting from the basion and passing through the pituitary tubercle, saw the amplitude of its rotations to constitute the movements of the malar.

15.3.2. *FDO orthopedists and orthodontists*

The 1950s also saw the emergence of the innovative work of two clinicians, Delattre and Fenart (1960), who, through their contribution, demonstrated the relationships between the face and the base of the skull. Indeed, Delattre (1951) then Fenart (1966) based all their research on the evolution of the cephalic skeleton by a vestibular orientation method, known since the first works of Perez in 1922 and Girard in 1923 on vestibular craniology.

The first observation of size resulting from the practice of this vestibular method, oriented along the axis of the semicircular canals, first gave implications on the ontogenic pathway. The successive positions occupied by a given point during growth reflect the ontogenic path oriented from this point. Thus, the curves of the back of the skull obtained in humans rotate clockwise (the so-called positive rotation), whereas in all primates, it is rather the opposite that occurs (the so-called negative rotation) after a beginning identical to that of humans (Fenart 1970). It should be noted that the latter is a tendency to quadrupedalism. Concerning the face, these two authors show that the mechanical laws are reversed to those of the back of the skull, so that in humans, the curvatures converge backward and downward, whereas in Primates, they are oriented upward and forward. Many authors later found these same structural implications in human paleontology (Clarck Howell 1972; Dambricourt-Malassé 1987, 1993).

Other parameters of analysis were highlighted by Fenart (1970) in order to show the relationships between the bi-parietal sphere, the occipital hole and the triturating plane in humans and chimpanzees. These occipitofacial relations are objectified by the geometrical construction of a sphere, in which a radius starting from its center and passing through the middle of the occipital hole in humans touches the sphere at a point that is tangent to the direction of the grinding edge. The angle between this edge and the radius is generally to the right. In *Pan paniscus* , the surface of the sphere remains away from the grinding plane and the angle is very open. In other words, the cranial tilt leads to the relative reduction of the jaws during hominization.

According to Delattre and Fenart (1960), among the biodynamic movements that follow the phenomena of occipital rotation and sphenoidal angulation, we can cite: the ascent of the whole base, especially the pituitary gulch and the tip of the petrous bone with posterior rotation of the basilar angle, the lengthening of the zygomatic arches and the lateral parts of the orbits, the lengthening of the pterygoid apophyses, the increase in the height of the maxilla and the ascending mandibular ramus with lowering of the palatal plane and the mandibular basilar plane, the accentuation of the nasal spine and the chin, the withdrawal of the upper and lower alveolar dental arches, in relation to the maxillary and mandibular bases.

In the therapeutic field, orthodontists who are reputed to be involved in the biomechanics movement have highlighted a certain number of clinical parameters, especially in young children, before and after the eruption of the first permanent molar (known as the 6-year molars). The identification of different structural types in normal, abnormal or pathological individuals, or their diagnosis by facial typology, has been followed since the classifications of Angle (1899), Sassouni (1959), Rickettes (1960, 1975), Delaire *et al.* (1965), Château (1975), Delaire (1978, 1980), Langlade (1981), Enlow (1982) and Deshayes (1986, 1993, 1997).

All these authors have insisted on the relationship between the face and the different bones that make up the cranial base of cartilaginous origin. For Delaire (1980, author's translation):

> Facial morphogenesis is conditioned not only by the state of the skull (vault and base) and the direct influence of the skeletal parts on each other, but also by that of the soft tissues, also, since the anterior base of the skull constitutes "the construction field" under which the face is built, it is immediately obvious that any change in its dimensions and proportions modifies the state of the face in parallel.

For Deshayes (2000), the eruption of the first permanent tooth appears to be crucial for the relationship between the base of the skull and the face. Indeed, for this author, from a basic cranial dynamic of neural origin, we will pass to a facial dynamic with the eruption of the 6-year molar tooth and the impact of the oro-facial functions. The observations of this author in a present-day child reveal that the inclination of the occipito-pterygoid plane (Ob-Pti) is almost always positive until the exit of M1. At the time of this eruption or shortly after the age of 6 years, a phenomenon of reverse inclination appears, namely, a negative occipito-pterygoid plane with respect to the sphenoethmoidal planum (Deshayes 2000).

Thus, as early as 1986, Deshayes considered that dental problems go beyond the facial sphere alone and rather contribute to a more global problem of the skull. The introduction of the concept of craniofacial biodynamics thus integrates the notion of organization of the different skeletal parts according to the forces applied to them. However, in view of the mosaic of structural frameworks, this author recommends that the study of craniofacial contraction, as well as the sphenoidal angle, should be analyzed separately from the biomechanical process that performed the flexion, because if there are very flexed specimens favoring a class III type evolution and those little flexed determining class II, the fact remains that some structures do not perform the same scenario. This is particularly the case with discordances, in which certain bony parts of the base of the skull carry out inverse or autonomous kinetics in relation to neighboring bones (Deshayes 2000). These particular craniofacial dysmorphoses escape the generally accepted notion of concordant dysmorphoses in craniofacial orthopedics. Prehistoric and historical populations sometimes present the same discordant structural scenario, whose kinetics of flexion and extension are inversely related to facial mechanics (Hadjouis 2001, 2003d, 2006a, 2006b, 2006c).

15.3.3. *Osteopaths*

The principles of osteopathy, which are not new, integrate the basi-cranial notion not only in the physiological and mechanical movements of the spheno-occipital joint and the beveled edges, but above all the understanding of the skeleton in its postural globality. Indeed, as early as 1874, Still described the general osteopathic theory by emphasizing the link that exists between visceral pathologies and disturbed vertebral structures. But it is especially with Sutherland (1939), a former student of the American School of Osteopathy (Kirkville, Missouri), that the notion of craniofacial bone mobility takes all its importance, especially with the sphenoid bone, which is according to him, the primordial element of craniofacial biodynamics. The description of cranial sutures, the shape of the margins, bevels and their inversion is presented in his book *The Cranial Bowl*. The major principle of Sutherland's work and his successors is summed up in the specific movement that is achieved from specific axes for each of the bones of the vault and base and that the movement of one bone would induce that of the other bones under a primary guiding force, in this case the variations in pressure of the cerebrospinal fluid.

More modern biomechanical authors emphasize the physiological movements of the spheno-occipital joint and the processes of flexion and extension. Thus, the mobility of the parts in relation to each other (temporal, occipital sphenoid) is dependent on the proper functioning of the peripheral parts (Altiéri 1984). According to Altiéri, the cranial biomechanical system is a spheno-occipital joint of great importance since it ensures the maintenance of craniofacial balance. However, Sutherland (1939), Upledger (1983), Clauzade and Daraillans (1989, 1992) and Altiéri (1984) insist on the factors that disturb the normal movement of the base of the skull as well as the growth rhythms. Among them are craniofacial dysmorphoses (blocked movements in flexion or extension), torsions, or right or left lateral rotational flexion. All these basi-cranial lesions induce facial deformities that destabilize facial harmony.

15.3.4. *Recent work in human paleontology and paleoanthropology*

15.3.4.1. *Spheno-occipital and neural dynamics in Old World fossils*

Based on the flexion of the base of the skull of Primates, Dambricourt-Malassé (1987, 1993, 2004, 2006) proposes a thesis that translates into an initially embryonic dynamic obeying the phenomenon of craniofacial contraction. In fact, this is no longer post-natal in *Hominidae*, but endocranial, preserved until the adult stage.

In order to understand the mechanisms that led to this tilting of the base of the skull, the author was first inspired by Gudin's pantograph, a geometrical construction in profile. Another transverse pantograph was exploited to take into account the future mandibular arch and the orientation of the bones of the base and the cerebral skull. The same general law of biomechanics and dentofacial orthopedics is found here, namely, "the more the bones of the face and the two jaws are slid under the forehead and widened, the more bone tissue in front of the occipital hole is straightened". However, the origin of the flexion and activity of spheno-occipital synchondrosis is now sought in embryogenesis by means of an angle in order to clarify the possible dissociation between the basi-occipital and sphenoidal vectors (Dambricourt-Malassé 2004).

The originality of this research consists of, since his thesis of 1987, the links between the straightening of the internal base, the development of the nervous system and the occlusal posture of *H. sapiens* are of embryonic origin. This research joins the Saltationist concept on the one hand, and the kinematic and occlusal concept on the other hand, all within an ontogenetic, phylogenetic dynamic (ante-*Homo*, *Homo* and Panins taxa). Following the discovery of the embryonic origins of these relationships, these concepts were developed and revitalized by Anne Dambricourt on the basis of the fundamentals acquired through medical imaging, DFO and orthodontics, whose initiators of craniofacial dynamics were Gudin, Delaire and Deshayes, and of the dynamics of craniofacial relationships by Delattre and Fenart. Since 1987, the date of her doctoral thesis on *L'ontogenèse, la paléontogenèse et la phylogenèse du corps mandibulaire catarhinien* (Ontogenesis, paleontogenesis and phylogenesis of the Catarhinian mandibular body), and *Équilibres bipèdes permanents, origine embryonnaire, morphogenèse, équilibre occluso-postural, conséquences pour l'évolution psychomotrice et comportementale des Hominidés* (Permanent bipedal equilibria, embryonic origin, morphogenesis, occluso-postural equilibrium, consequences for the psychomotor and behavioral evolution of *Hominidae*), as part of her HDR supported in 2011, she demonstrated the internal mechanisms that change from embryo to adult through comparative anatomy (chimpanzee, gorilla and orangutan), but including homeobox genes in the context of the Evo-Devo paradigm. Also Anne Dambricourt and the co-authors who reconstructed the base of Sinanthropus III (770 ky), compared the CT scans of 55 living *H. sapiens* of both sexes, adolescents and adults, 41 chimpanzees of increasing age, three Middle Pleistocene *Homo erectus* of Java, five *Homo neanderthalensis* and two Upper Pleistocene *H. sapiens*, and applied a new biometric protocol to them in order to compare the trajectories of chordal and pre-chordal embryonic dynamics. The results confirm the interest of applying this protocol to fossils in order to clarify taxonomic interpretations and evolutionary modalities between gradualism and saltationism.

15.3.4.2. *Craniofacial and occlusal dynamics of anatomically modern humans in North Africa*

In paleoanthropology, the larger size of the *Sapiens* populations of the Upper Paleolithic and those from more recent periods has allowed a meticulous structural teleradiographic analysis of each individual (Hadjouis 2003d, 2014, 2015a, 2015b, 2016a). Thus, the structural types of the Ibero-Maurusians and other North African settlements (Mechta-Afalou, Mechtoïdes and Proto-Mediterraneans) have not only shown concordant dysmorphoses of the flexion or extension type, but also discordant dysmorphoses, as well as a large number of basic cranial, arch and facial asymmetries, some of which are associated with spinal deformities (Hadjouis 1998b, 1999, 2000a, 2000b, 2001, 2002, 2003d, 2006c, 2008b, 2011). Biodynamic morphogenesis of the different bone parts reveals a close relationship between the base of the skull and the face. Dysmorphoses and asymmetries are, for the most part, induced by a basic cranial lesion, the causes of which are diverse (Hadjouis 2014).

The examples of pathological skulls, found with complete skeletons, show the association of lesions of a congenital torticollis for example with infracranial regions (cervical spine, scapular belt) (Hadjouis 1998b, 2006a, 2006b, 2006c).

The work of young researchers in human paleontology and odontology shows the same results as their predecessors in the craniofacial field (Carré *et al.* 2006) (Aoudia-Chouakri 2003; Cho 2005; Korpal 2005).

16

Identification of Malformation

16.1. Craniostenosis, a history of sutures

The literature on this cranial malformation is extensive. It has been identified and widely described both in archeological collections and in contemporary individuals under medical supervision in neurology. However, the repercussions of certain types of craniostenosis on the face, the occlusion and sometimes the cervical spine are rarely brought to light. The intimate association between plagiocephaly, one of the forms of craniostenosis, and its asymmetrical consequences proves the fragility of the links in the craniofacial puzzle during a sutural injury, in this case the injury of the coronal suture for this specific case.

The development of sutures of the human skull and of all mammals has long been at the heart of research carried out by anatomists, historians and physicians, from Herodotus, through Hippocrates and his followers, but also Celsus, Aristotle, Vesalius and Hunault. Also as early as 1547, it could be seen that in the disharmonious growth of a child's skull, the early closure of a suture could be responsible, because the analysis of the sutures and its obliterations by Virchow, Welker, Sauvage, Broca, Todd, Lyon and Vallois was also established according to a specimen estimating the age of death. Later, Bocquet-Appel and Masset (1977) based their research on this field within the framework of paleodemographic studies of Neolithic populations, using "probability vector" methods in the estimation of age by endocranial and exocranial sutures of both sexes.

Among the most common forms of craniostenosis are scaphocephaly (premature welding of the sagittal suture, lengthening the skull and reducing its width) (Figure 16.1), trigonocephaly (premature welding of the metopic suture giving the skull a triangular shape, bringing the eyes closer together), plagiocephaly (early and unilateral closing of a coronal suture), brachycephaly (premature welding of the coronal sutures giving the skull an excessive enlargement) and acrocephaly (damage to the coronal and sagittal sutures). The small size of the skull and the roundness of its contours confuse it with a microcephalic skull. Oxycephaly (coronal closing associated with the closure of the sagittal and lambdoid sutures) is the "sugar loaf skull". Pachycephaly is very rare; it causes early synostosis of the lambdoid sutures and gives the skull a short, flattened morphology which is confused with positional flattening. Children with craniostenosis are followed medically, as the malformation presents many complications starting with intracranial hypertension. The most serious forms are accompanied by mental disability, visual disorders and exophthalmos with sometimes complete blindness due to optic atrophy (Maroteaux 1995).

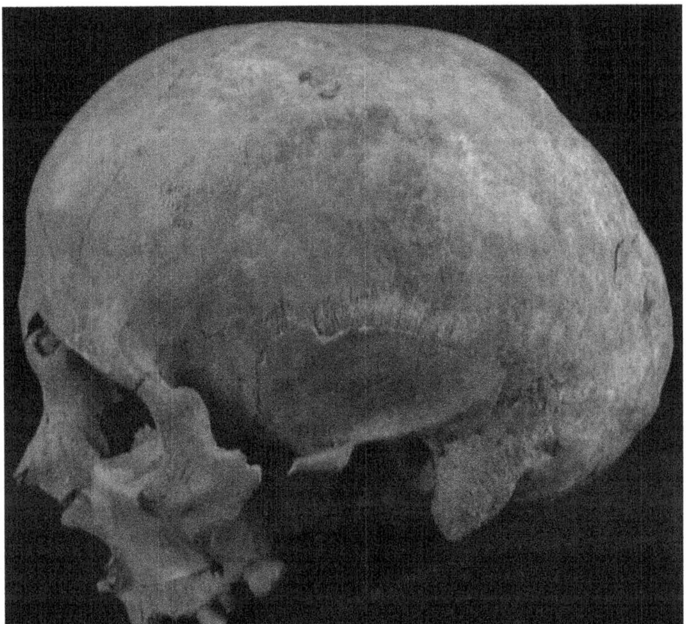

Figure 16.1. *Skull in left lateral view from the medieval necropolis of the Saint-Nicolas church in La Queue-en-Brie showing scaphocephaly due to early closure of corono-sagittal sutures (© Allard)*

Plagiocephaly in particular, followed by scaphocephaly (Figure 16.1), causes the most damage to craniofacial morphology. The craniostenosis of the archeological populations we studied, some in collaboration with Dr. Katz (Carré *et al.* 2006), were compared with the genetically determined dysostoses that the latter had been analyzing in children for many years (Hadjouis and Katz 1998; Hadjouis 1999, 2000b, 2012b, 2018). Clearly, in anthropology as in paleopathology, there are no bony criteria to guide us toward this type of diagnosis in order to make the link with the forms of neurological complications that contemporary medicine is aware of, and we will not take this step. However, it is useful to remember that patients, most often newborns, who present these types of craniostenosis, their radiological examination allows us to precisely locate the synostosis and to specify its extent. Some forms are rapidly progressive, and successive examinations show their rapid extension. Radiological interpretation shows a closed basal angle in longitudinal craniostenosis, but is normal in coronal damage. In the latter, the orbital angle (roof angle of the orbit and of the clivus) remains very open, the roof of the orbit becoming vertical. Digital impressions corresponding to the pressure of the cerebral convolutions on the internal specimen are also very marked.

The mode of genetic transmission is not clearly established and the frequency of this type of disease appears to be significantly higher in boys. While some authors suggest a dominant heredity, others suggest a recessive condition. Even if the genesis of these malformations remains mysterious, it should be noted that in the case of isolated craniostenosis, rapid surgical treatment excludes any mental disability (Hadjouis and Katz 1998).

All the skulls analyzed with true plagiocephaly and no positioning indicated the total absence of one of the two coronal sutures, followed by significant asymmetry of the vault, face and base of the skull. At the vault, the anterior movements of the hemi-skulls (as opposed to those with early closure of the coronal suture) were accompanied by a more developed frontal lump, while the opposite sides showed inverted features, including a flattening of the frontal bone. In other specimens affected by this deformation, in addition to the vault, the asymmetry concerned the entire basi-cranial and palatal complex (occipital scales, temporal scales, temporo-mandibular joint (TMJ) mastoid apophyses, petrous pyramids, basi-sphenoid, zygomatic arches and maxilla).

Concerning scaphocephaly, in addition to the premature closure of the sagittal suture, some skulls developed mixed scaphocephaly which has become generalized to the sagittal, coronal, temporoparietal, lambdoid, zygomatic sutures, zygomatico-maxillary, giving the skull an ultra dolychocephalic shape characterized by antero-posterior elongation, flattening of the lateral walls, heightening, platybasia, depression of the TMJ and excessive development of the mastoid apophyses in response to the depression of the base of the skull.

The medical literature, particularly that devoted to children's diseases, reports a large number of craniofacial syndromes and dysostoses of genetic origin that also associate various disorders (mental disability, eye and dental disorders, aplasia of certain bone elements, etc.).

16.2. Craniofacial asymmetries

It was not the deforming morphology of the skull and the face that was of prime importance, but rather its consequences on the loss of facial symmetry and occlusal balance, their repercussions on the underlying parts of the base of the skull being able to be generalized. Three types of asymmetries were selected: asymmetries of the vault, those of the face and those of the base of the skull. The diagnosis of these malformations was established according to the criteria of dentofacial orthopedics and orthodontics (Langlade 1981). These were based on several planes, the most significant of which are: the supraorbital plane, which allows the symmetry of an orbit to be determined in relation to its counterpart, and the mastoid plane or basic cranial plane where the sternocleidomastoid muscles are inserted, which allow for balance. In the case of a loss of equilibrium with respect to this plane, the etiopathogenic specimen, despite its variety, suggests retrospective diagnostics such as facial torsion, vertical asymmetry by mandibular hypocondylar, sagittal asymmetry evoking a peripheral facial paralysis and many others. In the case of basic cranial asymmetries, it is the lesions of the base of the skull that affect all or part of the face, just as the TMJ will prove to be a true indicator of imbalance (orientation, depression, arthrosis, arthritis and dislocation) (Hadjouis 1999, 2018). Biometry and/or medical imaging of the base of the skull and palate are classical and allow a rapid determination of the imbalance by assessing the antero-posterior lengths on both sides in relation to the tangent of the prosthion-basion plane. However, it is above all the disjunction of the spheno-occipital synchondrosis that will be extremely useful, since it immediately directs the causes of the imbalance (Figure 16.2).

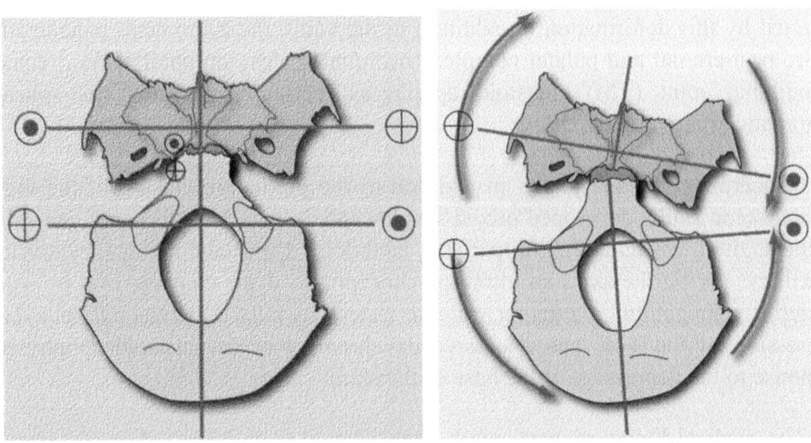

Figure 16.2. *The different forms of lesions of the spheno-occipital synchondrosis setting up an asymmetry of the base of the skull (rotation lateral flexion, torsion, plagiocephaly, torticollis, syndromes) (from Altiéri (1984) modified). For a color version of the figure, see www.iste.co.uk/hadjouis/skull.zip*

Apart from certain syndromes that unbalance the face, facial asymmetries most often result from malformations of the vault and base of the skull.

Of membranous origin, asymmetries of the vault, like craniostenosis and in particular plagiocephaly, are reflected on the face and the base of the skull. It is one of the rare forms of craniostenosis where facial torsions have given rise to new, descending mechanical functions, unbalancing the original symmetry of the mandibular condylar process. The TMJ, one of the main actors of sagittal dysharmoniacal kinetics, located in the center of the craniofacial puzzle at the vertical level, will in turn unbalance the temporal and the middle and lower levels of the face. The asymmetry of the temporals, which manifests itself by the advance of one of its injured poles (right or left), will be all the more decisive for the appreciation of this type of malformation, since it is from this anterior deviation, pushed by the new functions of the TMJ that all other unilateral bony elements of the base of the skull will follow (mastoid apophyses, occipital articular surfaces, jugular fossae, petrous pyramids, zygomatic arches and maxilla). In the end, it was found that the damage is of the same nature as if the origin was basi-cranial.

In addition to sagittal asymmetry, the vertical asymmetry of the face will also lead to other imbalances which result in a lowering of the orbits concerned, a more significant depression of one of the TMJs, a deformation of the mandibular condylar process, a lowering in height of the condylar process or of the vertical ramus of the mandible, a malocclusion (Figure 16.3). In total, true plagiocephaly affects the face both sagittally and vertically.

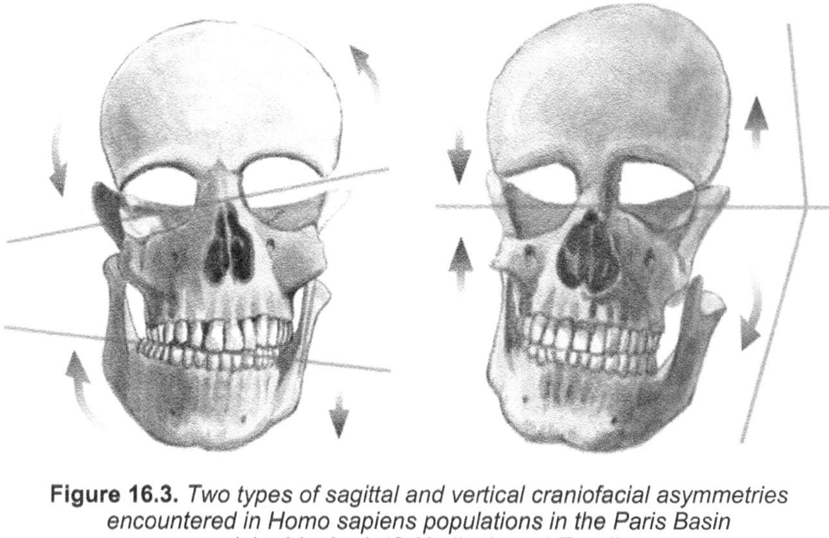

Figure 16.3. *Two types of sagittal and vertical craniofacial asymmetries encountered in Homo sapiens populations in the Paris Basin and the Maghreb (© Hadjouis and Tosello)*

The base of the skull is different from the vault by its cartilaginous nature and is intimately linked to the cervical spine. We will see further on that while humans have many craniofacial malformations and in particular asymmetries and other malocclusions, it is mainly due to their adaptation to standing upright since the straightening of the body. The quadruped adaptation of other vertebrates has not experienced the structural imbalances of the skull and human face precisely because of a perfectly balanced posture thanks to the arrangement of the centers of gravity in a horizontal cephalo-caudal axis.

The arrangement of the centers of gravity in a vertical cephalopodal axis that human beings experienced during postural adaptation on the lower limbs has made things more complex and the downward mechanical load has become more pronounced. Not only is the tendon-ligament-nervous system modified in its topography, like the pinching of spinal nerves in the supporting regions such as the lumbo-sacral region, but all the even parts of the standing position will experience on one of their right or left sides descending and ascending asymmetries with innumerable causes. Likewise, the anterior-posterior side will be modified according to an irregular adaptation between the forefoot, the middle foot and the rearfoot.

16.2.1. *Examples of craniofacial asymmetries*

Among the most commonly found asymmetries are lateral flexion rotation, plagiocephaly or pseudo-plagiocephaly, or lateral strain.

– *Lateral flexion rotation*: the effects of this structural disturbance have an obscure or unknown cause. It may have a peri-natal, congenital or acquired origin during childhood or adolescence. Malfunctions in the organization of the skull and face during growth are found both at the vault level through a sutural disturbance, at the cranial base, affecting the sphenoid, the basi-occipital or the spheno-occipital synchondrosis. Faced with such asymmetries, the face is often, if not exclusively, affected by a basic cranial origin (Figures 16.4 and 16.5).

From above, the vault shows a bilateral asymmetry (hemi-vault moving forward and the other half moving backward). The two opposite movements develop on the anterior side a protrusion of the supraorbital region and the frontal hump, while on the lateral side, the development of the parietal hump appears to be less and staggered. On the opposite side, the supraorbital region and the frontal hump are erased and receded, making the parietal hump recede in its movement.

In facial view, the asymmetry is double in its sagittal and vertical axes. In this type of malformation, there is a torsion of the face with a lowering of one of the hemi-faces reaching the middle and upper levels. The median is broken giving rise to an offset maxillary symphysis. The features of this double asymmetry are even more demonstrative in basal view. This type of rotation acts on the base of the skull by unilateral rupture of the spheno-occipital synchondrosis. On one side, this anteriorizes the large wing of the sphenoid, giving rise to a frontalized cheekbone with a maxilla in internal rotation and a sphenoidal hemibody situated low down, while on the opposite side, the maxilla is in external rotation developing a vertically more elongated face (Figure 16.4).

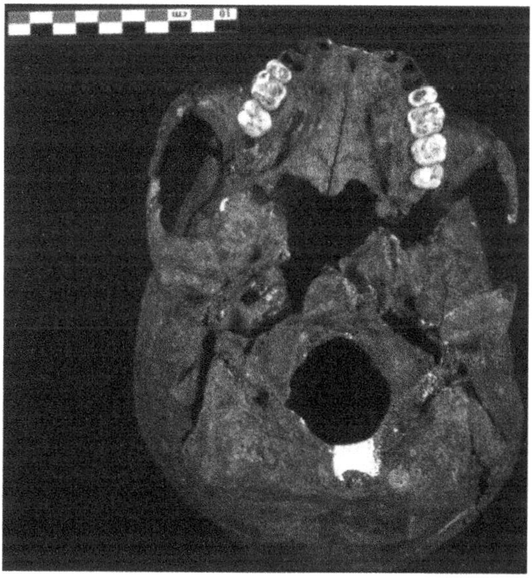

Figure 16.4. *Skull seen from below of a Homo sapiens from the Upper Paleolithic from Afalou (Algeria) showing significant basic cranial asymmetry with rupture of the spheno-occipital synchondrosis followed by left rotation (right in the photo) from the temporal bone (mastoid apophyses, TMJ, zygomatic arches, malar, maxilla). The result is a torsion of the face (© Hadjouis)*

– *Plagiocephaly or pseudo-plagiocephaly*: this example of malformation is mentioned in the chapter on craniostenosis and shows that the cranial deformity is linked to the only premature synostosis of one of the coronal sutures. Often associated with neurological disease in children, its asymmetry is disabling and unsightly giving the face an exaggerated sagittal torsion. The other related form, pseudo-plagiocephaly, so called because of the absence of any sutural damage, has developed the same deformities.

– *Lateral strain*: this terminology borrowed from osteopathy concerns a lesion or rupture affecting the spheno-occipital synchondrosis. The lateral strain type deformity acts on the SOS through a rupture when it is well conserved. The result is a lesion that predisposes the basi-sphenoid and basi-occipital in an inverted lateral position (Figures 16.5 and 16.6). Thus, the median axis being broken, the main bones of the base of the skull are out of axis putting the occipital in one orientation, the sphenoethmoid-frontal complex in another, advancing the orbit pushed by the hemi-frontalization of the wing of the sphenoid. The mandibular body is laterally offset. The skull is then inscribed in a parallelogram (Deshayes 2001, Cranexplo).

– *Asymmetry by peripheral facial paralysis*: the example of this malformation is represented by two subjects from the Merovingian and Carolingian necropolises of the Val-de-Marne department. One, whose symptoms are perfectly preserved, shows a strong asymmetry of tartar deposition by salivary flow on the left maxillo-mandibular jugal teeth. Associated with this paralysis are malformations on the right side, in particular the reduction in height of the maxilla and the zygomatic arch. In terms of the base of the skull, the lateral flexion rotation caused all the left parts to advance, in addition to the flattening of the hemioccipital: occipital articular surface, mastoid process, TMJ and zygomatic arch. Similarly, there was strong erosion of the articular tuberosity of the temporal bone, a sign of unilateral mandibular dislocation (Figures 16.7–16.9).

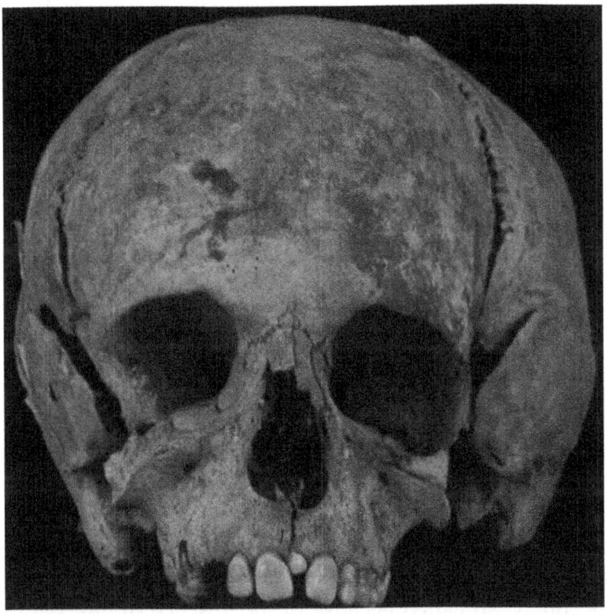

Figure 16.5. *Child's skull in facial view from the necropolis of the Saint-Nicolas church in La Queue-en-Brie showing significant facial asymmetry (torsion of the face and tilting of the left hemi-face). Note the presence of a mesiodens, a tiny supernumerary tooth, between the two central incisors (© Barrau)*

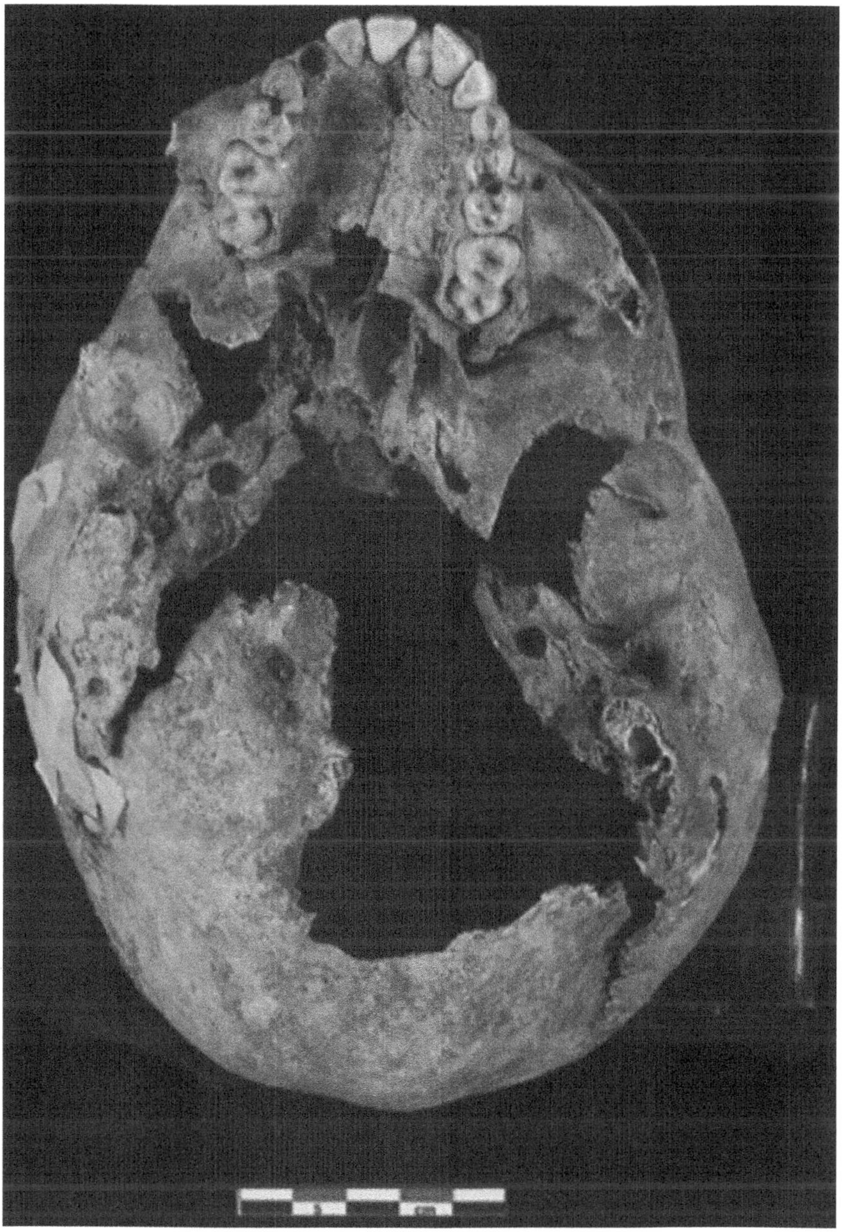

Figure 16.6. *Same individual, seen from below. Note that the banana shape of the asymmetry is even more impressive at the base of the skull and the lower parts of the face (© Barrau)*

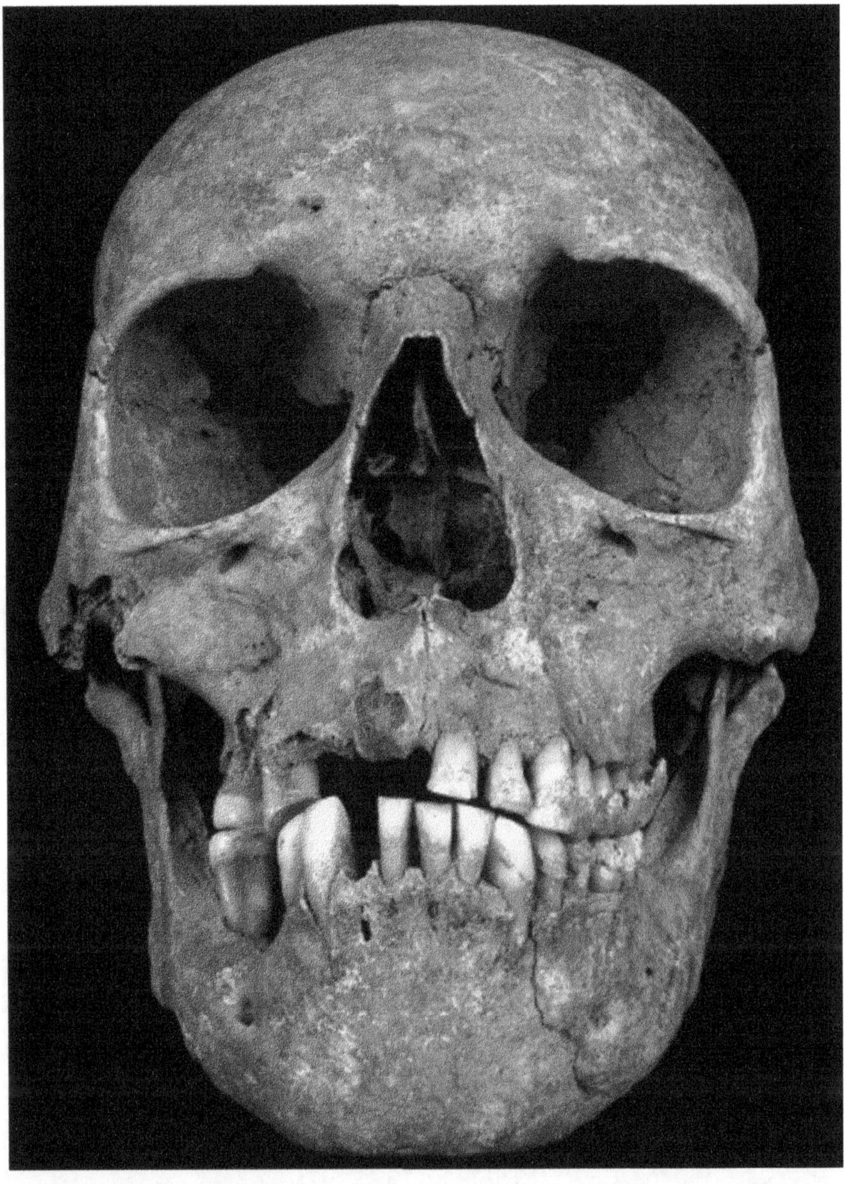

Figure 16.7. *Facial asymmetry suggestive of peripheral facial paralysis with respect to a strong asymmetry of tartar deposition by salivary flow on the left maxillo-mandibular jugal teeth. Associated with this paralysis are malformations on the right side, in particular the reduction in height of the maxilla and the zygomatic arch, Merovingian necropolis of Saint-Christophe de Créteil (© Allard)*

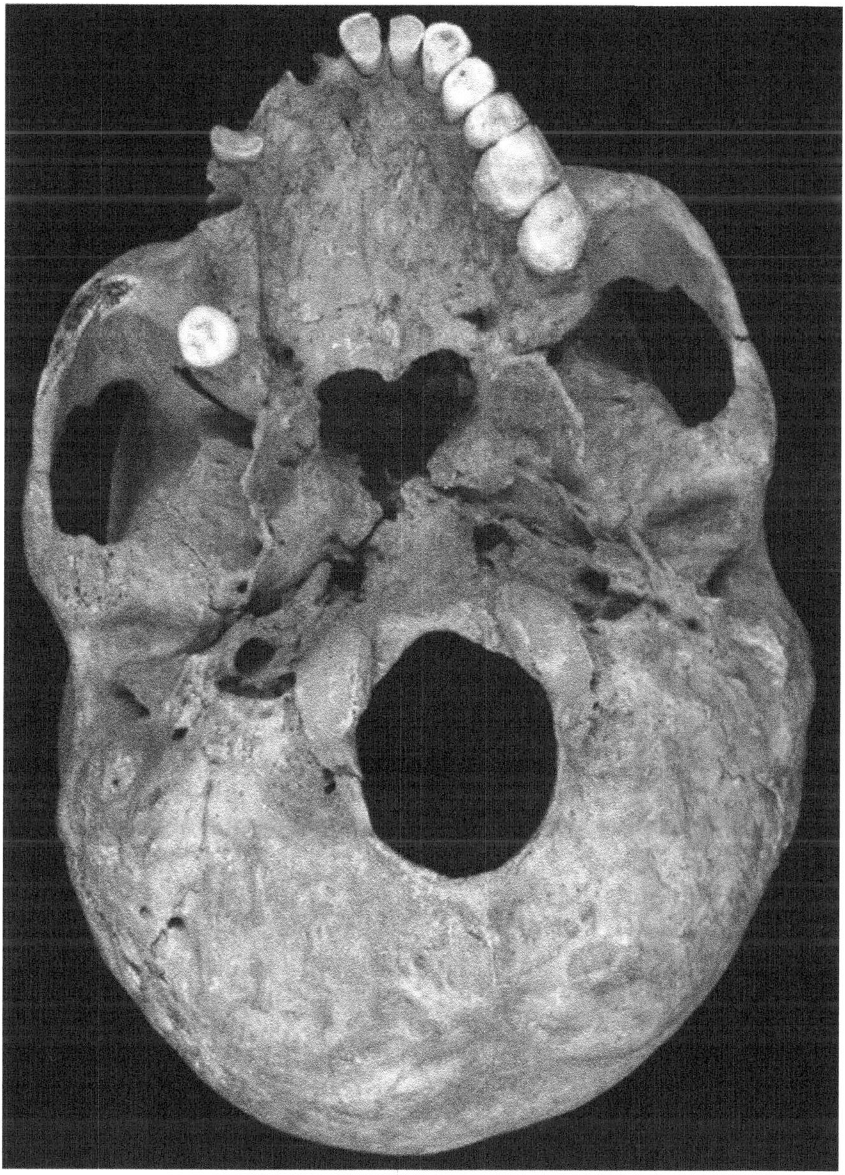

Figure 16.8. *Base of the skull of the individual with peripheral facial paralysis showing strong asymmetry. Lateral flexion rotation has caused all the left parts to advance, in addition to flattening the hemioccipital (occipital articular surface, mastoid process, TMJ and zygomatic arch). Similarly, note the strong erosion of the articular tuberosity of the temporal bone, a sign of unilateral mandibular dislocation (© Allard)*

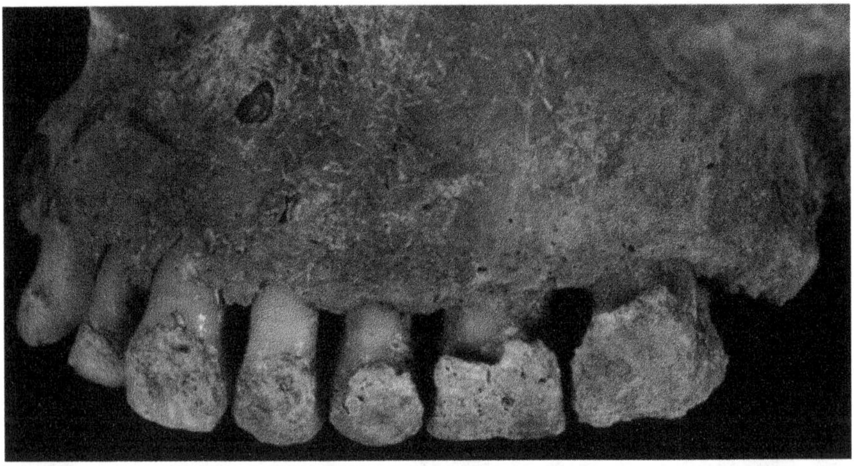

Figure 16.9. *Large unilateral calcareous deposit produced by salivary flow following peripheral facial paralysis in a man, Merovingian necropolis of Saint-Christophe de Créteil (© Allard)*

16.2.2. *The importance of the spine and its effects in basic cranial equilibrium or disequilibrium*

During the verticalization of the human body, several physiological spinal curves were formed for postural balance and gravity adaptation to the ground through a bipodal position. The S-shaped spinal morphology forms three curvatures: a cervical lordosis with posterior concavity, maximum in C5, a dorsal cyphosis with posterior convexity, maximum in T8, and a lumbar lordosis with posterior concavity, maximum in L3-L4. When the body is straightened, two anatomical regions of the spine prove to be of prime importance for verticalization, but their instability also exposes them to the worst. At the top of the spine, the first cervical vertebra, the atlas, is responsible for the articular stability of the base of the skull while maintaining the necessary movements for the entire head. The slightest articular imbalance of the occipito-atlasic hinge will affect the entire base of the skull, sometimes even influencing the skeletal bases of the occlusion. A C1-C2 congenital block with unilateral welding can have harmful consequences on the balance of the base of the skull.

At the base of the spine, the lumbar segment and the lumbosacral hinge have the same mission of stability of the paired parts of the lower limb, but such a role seems even more acrobatic, since in addition to maintaining the balance of forces, the distribution of the latter is equal to the gravity load received. It is thanks to the formation of the lumbar lordosis when standing that the load-bearing parts, like the pelvis, are able to play their role. A distribution of the centers of gravity, which differs in Great Apes with a single-curved spine and posterior convexity, gives a gravity load located higher in the thoracic region, reflecting a semi-erect quadruped posture.

At the slightest imbalance of the spine, the destabilization of the central and odd skeletal element will have consequences on the above and underlying parts: scoliosis or low cypho-scoliosis unbalance, depending on their intensity, the upper limb by a dimensional asymmetry (limb shorter than its counterpart), a probable dislocation of the shoulder or secondary osteoarthritis. In the same way, the lower limb is also affected by the same effects, even more so given the gravity load (dimensional asymmetry, probable dislocation of the hip and secondary osteoarthritis). The foot can also be affected. The effects on the lumbosacral hinge are manifested by an asymmetry of the first sacral vertebra (S1) sometimes forming a unilateral lumbosacralization of S1. The effect can sometimes be the cause, or a congenital anomaly of the lumbosacral hinge (S1), and will cause a scoliosis with ascending and descending repercussions.

16.3. Psalidodontia or labidodontia?

The evolutionary mechanisms of *Sapiens*, introduced between 200,000 and 300,000 years ago in East Africa (Bodo in Ethiopia) and North Africa (Djebel Irhoud in Morocco), have enabled new organizational plans to be put in place. (within the genus *Homo*) which manifested themselves by neuronal complexification, a greater cerebral capacity and a repositioning of the cerebellum under the occipital, allowing a new anatomical interpretation of the growth trajectories thanks to this new *Sapiens* level. On the membranous and cartilagino-osseous level, the trajectories of craniofacial rotation have introduced new dynamics which are presented as follows: elevation of the vault and development of the forehead following the basic cranial flexion which allowed the architectural triangle to close, anterior tilting and rotation of the occipital scale, with advance of the lower occipital and upward movement of the lower sphenoid. This spheno-occipital rotation allowed the clivus to become a little more vertical thanks to the closing of the sphenoidal angle. The facial skull has become clearly more important than the spinal skull in the anterior-posterior direction. Orthognatia and psalidodontia were put in place. For the first time in the evolutionary history of the genus *Homo*, the *Sapiens* species was provided with a more straightened forehead, chin and an overlapping of the upper incisors over the lower ones.

The malocclusion and dysmorphoses of the juvenile populations of contemporary humanity are constantly progressing and these, whether they come from Europe, America, Asia or Africa, will require more and more orthodontic treatment. The causes are multiple and should be researched on a large scale. To this end, a debate has been underway for several years between orthodontists and paleoanthropologists on the period of appearance of the overlap and overjet of the anterior teeth of the maxilla.

Due to cuspid interlocking (presence of a large canine and interdental diastemas) in Great Apes and craniofacial structural kinetics of the flexion type (strong anterior rotation of the squamous part of the occipital bone), coupled with strong abrasion in *Homininae*, the occlusion is of the labidodental or end-to-end incisor type. Thus, even if only one jaw (maxillary or mandibular) was found during archeological excavations, it is easy to identify the function of the anterior guide function based on the dental wear of the occlusal surfaces of the incisivo-canine group. The anatomically modern populations, including contemporaries, on the other hand, are in supraocclusion or psalidodental (covering of the upper incisors and overjet). However, eating habits and in addition abrasive food has somewhat modified the dental joint, making it episodically labidodental after advanced tooth wear (Figure 16.10). Other functional mechanisms of the teeth and facial musculatures have modified this joint.

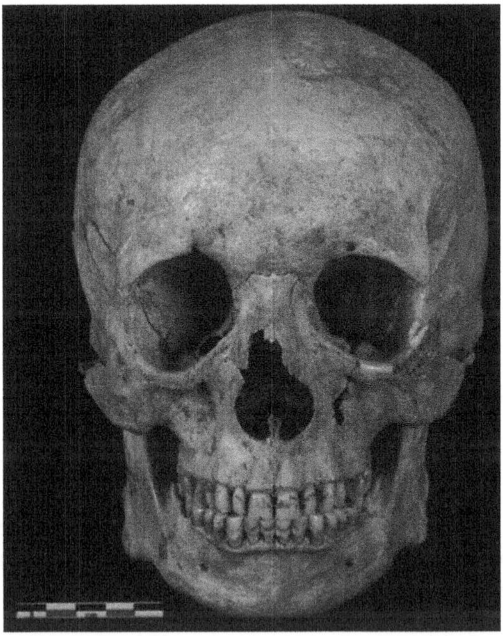

Figure 16.10. *Male skull in facial view from the necropolis of the church of Sainte-Colombe in Chevilly-Larue showing an incisive end-to-end dental bite in response to a class III structural specimen (© Hadjouis)*

Faced with an observation as complex as that of psalidodontia in contemporary populations, apart from dysmorphoses, it is essential to resort to the methods that manage the fundamentals of occlusion and even odontology in general. Occlusion, Garbarino (1979) tells us, is a position and not a state of rest that can be under variable pressure. The teeth are then subjected to forces. A change in function, an increase in the occlusal load and a change in the direction of the force determine the disorders of the structure of the overlying bone. The functional friction of dental articulation is of particular interest in the joint. This means that in the same dento-maxillo-mandibular position (in a closed mouth without movement), occlusion and dental articulation return to the same. It is therefore important in this analysis of occlusion and dental articulation knowing that the latter intervenes mainly in the masticatory system.

In addition, the anterior guide function of the maxillo-mandibular groups is a current reference that manages the incisivo-canine functions in all directions (anterior, lateral and anterolateral) as long as the incisors are the first to appear at the age of 6 months. It is in fact the incisor or incisivo-canine guide that plays a decisive role in the oro-facial functionalities and we will see further on, that in ancient populations, its presence or absence also intervenes in the craniofacial structural kinetics starting with the masticatory system.

In fossil or sub-actual populations, the anterior guidance is distorted by several phenomena whose functionalities, sometimes different, lead to the same results. The only drawback in this type of analysis is that the mechanics is often poorly distinguished in its primary function. In fact, it must be noted that the labidontic (incisive end-to-end) occlusion of Panins (*Pan troglodytes*), Gorillas (*Gorilla gorilla*) and Pongids (*Pongo pygmaeus*) shows an incisivo-canine sector entirely locked by the great development of the canine teeth and their reciprocal position in a diastema designed for this purpose when the mouth is closed. Concerning the *Homininae*, from *Ardipithecus ramidu* , to *Homo sapiens*, via the *Australopithecus*, *Homo habilis*, *Homo erectus* and *Homo heidelbergensis*, the same phenomenon of labidodontia remains with a canine stop, but differently from the occlusal causes of the Great Apes. One of the explanations of this type of occlusion is in part the para-function, recognized in any case in Neanderthals. This same para-function has been used in sub-actual (Middle Ages) or contemporary (Eskimos and Australians) populations. Trimming, polishing, spinning or tanning of hides are some of the many para-masticatory functions of these populations.

The crucial question that has arisen is whether labidodontia intervenes in the dental articulation of children and adolescents in anatomically modern and sub-actual fossil populations. After several years of research on several children's collections, the result is irrefutable: psalidodontia has been exclusively present in children since the Upper Paleolithic and none of these juveniles indicated a suspicion of labidodontia.

16.3.1. *The behavior of the dental articulation of juvenile Pleistocene and Holocene populations in the Maghreb and the Sahara*

In Anatomically Modern Humans (AMH) of North Africa (Afalou Bou Rhummel and Taforalt), the few individuals who did not have the upper incisors extracted were labidontic and most, if not all, would have had such an occlusion if they had not undergone this ritual. Indeed, comparative work carried out on the occlusion of humans from Mechta-Afalou and Taforalt (Chamla 1970; Hadjouis *et al.* 2000a; Hadjouis 2002, 2011; Hadjouis and Katz 2019), from Columnata (Chamla 1970; Hadjouis 2011) and the Hyena Cave and the Proto-Mediterranean from Khanguet-El-Mouhaad and Aïn-Dokkara shows the same development of the maxillo-mandibular anterior sectors. As the anterior guide function was missing after extraction of the upper incisors in young mecthoids, the masticatory system and facial muscle forces functioned differently. The skeletal bases were destabilized not by an effect of skull base flexion, but by a new mode of adaptation of the lower and middle stages of the face, which reconditioned the mandibular growth trajectory. The hypo-maxillary phenomenon produced by the removal of the upper incisors highlighted the development of a protrusive mandible, not solicited by the basic cranial flexion. The Spee's curve, exaggeratedly pushed by the lingual void, gave free rein to the development of the mandibular incisors and canines in height.

In addition, the juvenile series of Afalou (Afalou 4, Afalou 6 and Afalou 26) (Hadjouis 2003d, 2020c), Columnata (H21 and H43) (Chamla 1970; Hadjouis 2002, 2011), of the Neolithic Sahara from Tin-Hanakaten (child no. 5) (Heim *et al.* 2011; Hadjouis 2020c), Western Sahara (Petit-Maire 1979) were not incisive.

Indeed, in Afalou 4, this child of 3 years +/– 12 months had a complete lacteal dentition with M1 in their alveoli and was not disposed of any apparent malocclusion. The maxillo-mandibular set presented an alveolar-dental prognathism. The upper incisivo-canine block demonstrated a modern type of overlapping, different from the adult *Sapiens* occlusion (Figure 16.11).

Afalou 6 died at 18 months +/– 6 months, his lower left incisors and canine teeth fell out *postmortem*. The upper teeth present were I2, canines, D1, eruption of D2 in the alveolus, germs of M1 and germs of the permanent incisors. Even in the absence of the incisivo-canine areas of the mandible, an overjet of the upper incisors could easily be evoked, not far from 2 mm (required standard of overjet in contemporary humans).

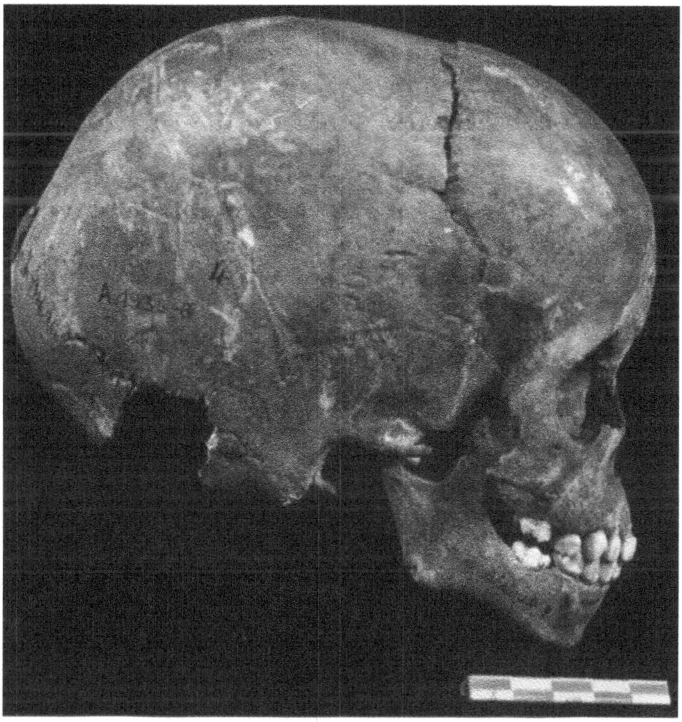

Figure 16.11. *Skull of Iberomaurusian child no. 4 from Afalou in right lateral view. The face shows alveolar-dental prognathism on a psalidodental joint. Homo sapiens children were the first to acquire a permanent psalidodental joint, unlike adults. In Neanderthals, children and adults had a permanent labidodental joint (© Hadjouis)*

Afalou 26 is even more interesting given his age of 12 years +/− 12 months. The mixed dentition of this tall child (or young teenager) fully justifies the question asked in this study. The mandible was not present, however, the alveolar protrusion of the upper incisors suggests a significant overjet. In addition, the rate of dental growth of this individual was significantly behind that of contemporary populations. Indeed, the teeth experienced a disturbance in the eruption of the milky and permanent teeth, in particular a delay in the dental breakthrough of the incisivo-canine sectors. The right deciduous canine had not yet fallen out, while the permanent canines were in the process of eruption. Likewise, the entire incisivo-canine group should have lost its temporary teeth a long time ago since the first two molars (M2 fallen out *postmortem)* came out and the M3 was in its socket.

In the end, this individual not only presented a considerable delay of the anterior teeth, but also suggested a psalidodental dentition.

In Columnata's mechtoid collection, an important series of infants and children makes up this population. Among the juvenile individuals, two complete skulls with their teeth are present and described by Chamla (1970). H21 and H43 are two children aged 8–10 years old and 12–15 years old, respectively. The upper incisors of the first child were overlapping and overhanging. On the other hand, in the second individual, the presence of dysmorphosis (class III) predisposed the lower teeth to mesiocclusion (lower canine and M1 located mesially in relation to the maxillary antagonists).

The Neolithics of Tin-Hanakaten delivered a complete skeleton of a child (*Homo* no. 5) perfectly preserved dated 7,900 ± 120 years (GIF: 5,857). The 4- to 5-year old child with Negroid characteristics had preserved part of his tegumentary envelope consisting of a reddish crust, some flaps of which still adhered to the periosteum of the skull and limb bones (Heim *et al.* 2011).

In spite of temporal discordance and alveolar prognathism, the craniofacial structure and occlusion of this child describe a growth whose dento-skeletal balance was eventually little disturbed. In fact, in terms of occlusion, this child, who had not yet fully erupted his 6-year old tooth, was moving toward an overjet of the upper incisors, the dental overlapping already being sufficiently in place (Figure 16.12).

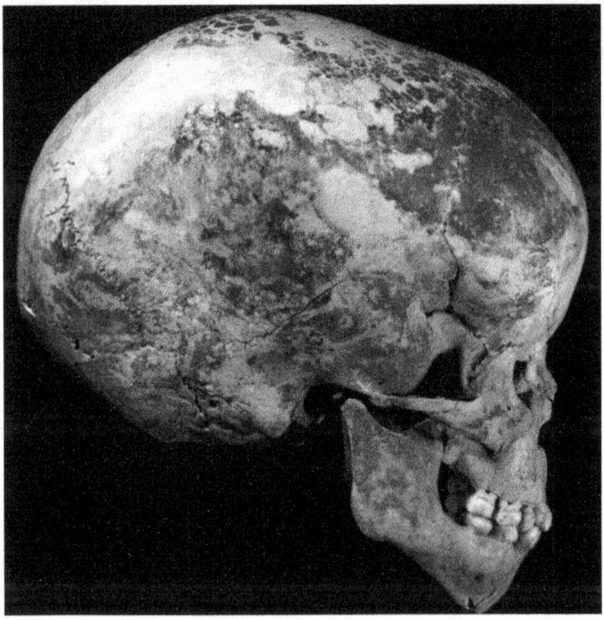

Figure 16.12. *Child's skull in right lateral view of the Neolithic of Tin-Hanakaten (Tassili, Algeria) showing a psalidodental joint (© Hadjouis)*

In the Holocene, this psalidodontia is noticed in all juvenile individuals with mixed dentition or permanent teeth. During the excavation of the human remains of Meniet (Ahaggar), Charon *et al.* (1974) made this clear. In Meniet IV (5,410 BP), a 12-year old child presented a psalidodental joint, but would probably have become labidont if the abrasion had continued, in view of chewing abrasive food.

The Holocene populations of the Atlantic Sahara studied by Petit-Maire and his team are somewhat oriented toward the same observation. Indeed, the dental analysis of Petit-Maire *et al.* (1979) of the Izriten necropolis (Cap Juby) (6,100 +/– 120 BP) showed in two mecthoid adult individuals with complete dentition, the presence of a bite called psali-labidont (almost end-to-end), while in a third aged 14 years old, the occlusion seemed to be labidont with respect to the advanced abrasion of the incisors.

In Sebkha d'Amtal (1,780 +/–100 BP but context dated 3,000 and 4,000 BP), the skull of a young child (Amtal 1) shows a psalidodental-like occlusion with a strong alveolar protrusion of the upper incisors (Petit-Maire and Fleichier 1979). Elsewhere, in the Sebkha Maharia (eastern Cape Barbas) (5,040 +/– 100 BP and 4,100 +/– 110 BP), this same team described among the burials found the skeleton of a young adult aged 20–25 years old with end-to-end occlusion.

In the important necropolises of Tintan and Chami (Quaternary Marine Sedimentary Basin of Mauritania) located not far from the bay of Nouadhibou for the former and behind Cape Tafarit for the latter, all adult individuals presented a labidodental joint while the young were slightly psalidodental, evolving rapidly with age toward labidodontia.

The conclusions of these authors on the occlusion of the mechtoid populations of the Neolithics of Mauritania suggest for the first time the passage from end-to-end occlusion to the current psalidodontia. These observations noted in symbiosis between paleooanthropologists and dentists (Petit-Maire and Charon 1979) on the occlusion of Saharan populations in the 1970s have never been noted by other authors.

This interesting question of the transition from end-to-end to incisal overlapping during the Saharan Neolithic was not necessarily shared with populations of other regions, and even more so with sub-current populations such as the European Middle Ages.

16.3.2. *Dental articulation and extraction of the incisors*

Dental articulation refers to masticatory forces or para-masticatory functions on the teeth of both maxillo-mandibular sets. The congenital absence or pulling of teeth from the incisivo-canine areas will certainly disrupt its function, the effects of which will be felt on the skeletal bases.

The phenomenon of dental mutilation of the maxillary or maxillo-mandibular incisors practiced among the populations of the Maghreb (especially in Algeria and Morocco) during the Upper Paleolithic and Holocene periods is well known to authors. However, while the ritual itself has been compared to those of certain tribes obeying the same rules of ritual practice, nothing has been observed, even less to an analysis of the occlusal behavior after incisor extraction. Although incisor extraction has concerned children and adults without distinction of gender, it has not been systematic to the general population, nor has it followed a regular pattern.

Among the skulls of the humans of Afalou analyzed from the series of Camille Arambourg from the Institute of Human Paleontology in Paris, 18 specimens had experienced extraction. The most common practice was the removal of the two upper central incisors (11 cases), followed by the removal of only one side, left or right (three cases, 2 I1 g., 1 I1 d.). Two individuals had the upper incisors removed on both sides of the maxilla (I1 + I2 g., I2 d.; I1 g., I1 + I2 d.). Similarly, in this collection, we find some individuals whose incisors were torn from the maxillo-mandibular set. Concerning removal in very young children, two individuals aged 5–6 years old had their upper left central incisor torn out in regard to temporary dentition. The good healing of the free edge of the pre-maxillary of one of them (Afalou 19) shows that removal had to be performed at an even younger age.

As previously pointed out, the practice of incisor extraction resulted in a modification of the Spee curve, which became exaggeratedly curved, as well as a rebalancing of the kinetics of the maxilla. Bone growth thus developed not only as a result of dental trauma paralyzing the dynamic functionality of the maxillary kinetics, but also at the same time developing an exaggerated height of the lower incisivo-canine row (Figure 16.13).

The hypomaxillia phenomenon produced by incisor extraction set in motion the development of pseudo-protrusive mandible, where the lower incisors protruded vertically and mesially without antagonistic contact. The Spee curve which defines an arcuate occlusion curve connecting the line from the vestibular cusp of the first mandibular premolar to the distal cusp of the second molar was exaggerated and did not seem to belong to any model among the fossilized hominids. In fact, the void in the anterior region of the maxilla required greater height development of the incisors and lower canines (Figure 16.14). Furthermore, it should be noted that even in the absence of upper incisors, the dental joint functioned despite this disability, in labidodontia like the majority of other adult individuals in this population.

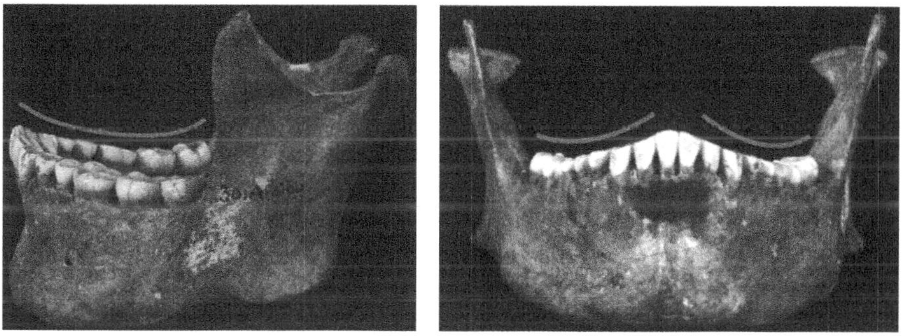

Figure 16.13. *Mandible in profile and face of Homo Sapiens from the Upper Paleolithic from Afalou (Algeria) showing an exaggerated Spee curve due to the strong thrust in height of the teeth of the incisivo-canine block due to the absence of the antagonistic teeth of the maxilla (© Hadjouis)*

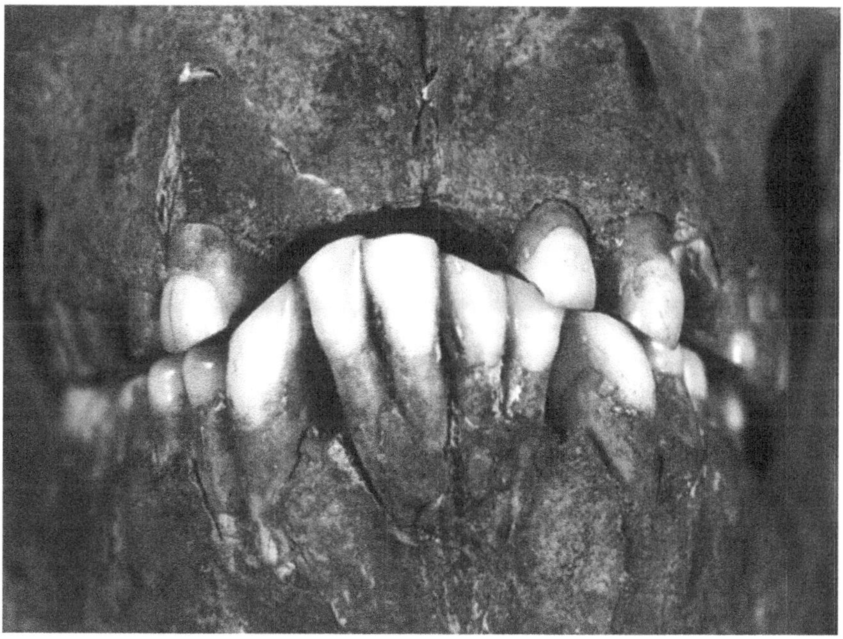

Figure 16.14. *Detail view of the extracted upper incisors and the development in height of the lower incisors on a man's skull from the Upper Paleolithic of Afalou (Algeria) (© Hadjouis)*

The other element to report, that is no less interesting, concerning mixed mutilations (upper and lower extraction) is the dental configuration of the lateral teeth that were close to those that have been pulled out. It must be said that although the pulling out of the upper teeth is clearly visible, both by the void produced and by the convincing healing, the extraction of the lower teeth is not recognized; it is almost unknown, due to the absence of the alveolar-dental void. The character is observed in three specimens, one of which represents the type species of Mechta-el-Arbi, the second a female Afalou skull and the third a Neolithic with the mechtoid characters of Oued Guettara (Hadjouis *et al.* 2000b; Hadjouis 2002, 2003). In Mechta-el-Arbi and the individual from Oued Guettara, early mutilation of the lower central incisors, which occurred during childhood between 8 and 10 years of age, reduced the alveolar-dental space at the symphysis level and contributed to alveolar erasure. Lingual rotation of the lateral incisors and canines helped this phenomenon. In this case, the mandibular teeth were removed from the permanent incisors from the age of 7–8 years old. In this particular case, the growth of the mandibular bone at the symphysis developed normally, without leaving a bone void at the location of the extracted teeth, helped by the support of the forces of the lateral teeth. In this aspect, the canine incisivo block took on the appearance of hypodontia, easily confused with a form of dental agenesis.

In the Proto-Mediterraneans of Algeria, the extraction generally affected both maxillo-mandibular sets and more particularly the eight incisors. For a long time, it was thought that this practice only concerned the female sex in the Proto-Mediterranean populations and that mixed practice was systematic. It is known today that some female subjects did not undergo any dental mutilation as shown by the skeletons from the shell mound of Bekkaria and Ain Bahir. In the same way that the male subjects from the shell mound of Aïn Dokkara and the site of Medjez 2 (skeletons 1 and 3) did not experience this mutilation either. On the other hand, male specimens from Ain Berriche had all eight incisors mutilated.

16.4. Para-masticatory functions of *Homo sapiens* in Algeria

The subjects that were analyzed concern three adult individuals from the sites of Afalou Rhummel, the Capsian site of Medjez II (El-Eulma, Sétif) and the Neolithic site of Oued Guettara (Murdjadjo, Oran). In all three cases, the function of the dental joint was unusual and did not follow a masticatory logic.

The first and most symptomatic case concerns a male individual aged between 25 and 35 years old found in 1967 during the excavations by the former CRAPE team on the Murdjadjo massif. In addition to the mechtoid features, the presence of extraction of the upper incisors and the extent of alveolar-dental lesions, the fossil also showed abnormal wear of the mandibular jugal teeth (Hadjouis *et al.* 2000b; Hadjouis 2010).

From the outset, both from the front and from the side, the imposing teeth of this Neolithic individual were surprisingly massive and sturdy, bringing it closer, like the whole of the skull, to the Paleolithic populations of Afalou (Figure 16.15). Indeed, the canine, premolar and molar sectors on both sides of the mandible demonstrated oblique wear on the vestibular side from the inside outward (Figure 16.16). The antagonistic teeth of the maxilla, on the other hand, showed normal horizontal chewing wear and did not appear to be related to this joint. Here, the lower wear occurred independently of the upper teeth, whose sole function may have only been para-masticatory. The importance of the bilateral wear shows that this para-function (not to be confused with the para-function known in orthodontics and FDO – digital suction, bruxism, etc.) was exerted for a long time on the masticatory system, which generated oblique wear in the premolar and molar sectors and acted like a tool making transverse movements. Examples of this kind of wear could be applied to the tanning of animal hides, which had to be done manually by applying strong pressure from left to right and downward. Curiously, the only teeth which have known this wear have developed periapical inflammatory lesions and periodontosis.

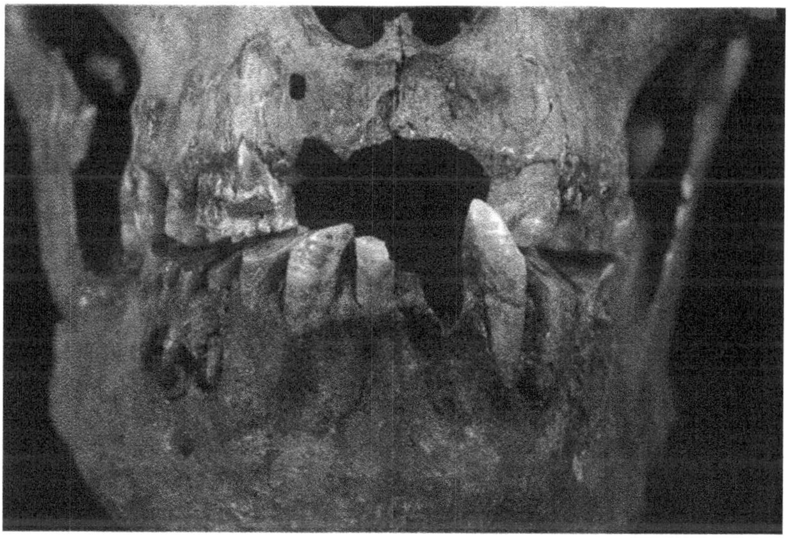

Figure 16.15. *Skull of a mechtoid man from Oued Guettara (Algeria) in facial view showing, in addition to extraction of the upper incisors, bilateral para-masticatory wear of the jugal teeth. The oblique wear from inside to outside is due to a para-function linked to a particular activity such as hide tanning (© Hadjouis)*

In Afalou Bou Rhummel's collection, there is an adult female individual (Afalou 33) whose second premolars and first molars of the maxilla show significant oblique wear on the lingual side, while the antagonists of the mandible show more classic occlusal wear, not responding to the superior dental obliquity. This type of wear results in an uninhabitable occlusion pattern, probably of para-mastricidal origin, whose vector (tool or other para-function) is difficult to suspect (Hadjouis 2002, 2011). Furthermore, it is interesting to note that the right lateral I2 that escaped total incisor extraction underwent a triple movement of gression, versioning and rotation that repositioned the tooth in alveolar protrusion at the location of the I1. The double vestibulo-distal wear of this incisor shows that it articulated end-to-end with the lower incisors despite mandibular mesiocclusion of the canine and M1.

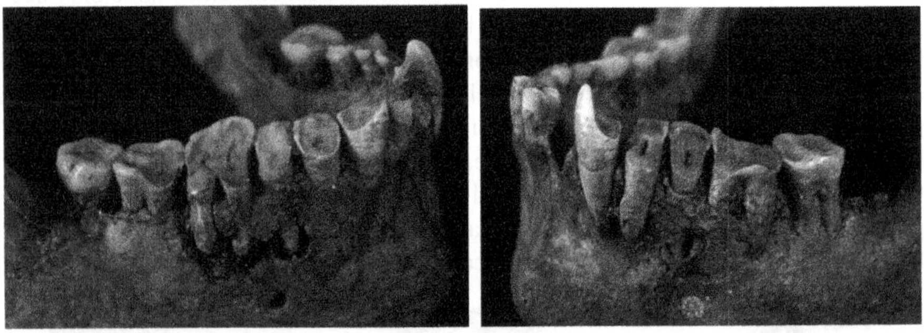

Figure 16.16. *Right and left mandibular jugal teeth of a mechtoid man from Oued Guettara (Algeria) showing details of para-masticatory wear (© Hadjouis)*

In the Capsian deposit of Medjez 2 (Fabrer-Camps 1975), a skeleton of an adult female (H2) was found who had an unnatural occlusion dentition. Indeed, the M1 and the second left premolar of the maxilla showed oblique wear on the lingual side in the same pattern as the Afalou fossil, while the corresponding lower teeth had horizontal wear. On the other hand, the upper incisivo-canine block presented a right I1 with lateral wear that could articulate with the lower incisors if they were still present, which was not the case. However, the author who described an extraction of two upper incisors (in reality only the left I1 was absent) noted that the double wear of the right I1 could be caused by voluntary trimming, which cannot be false. It should also be noted that the three skulls (H1, H2 and H3) presented a psa-labidodental bite.

As regards para-masticatory activities, they have existed both in fossil humans, especially Neanderthals, and in the Eskimo and Aboriginal populations of Australia. However, work activity is localized in the anterior sectors and is of particular interest to the incisivo-canine group of the maxillo-mandibular sets. This function has moreover given rise, as in a very abrasive mastication, to a labidodental joint. However, the whole question lies precisely on occlusion and masticatory functions in young people and adults. Important series, such as the ones we have just listed above, indisputably show the equation: young = absent or slight abrasion = psalidonts (psali-labidonts), and adult = strong abrasion = labidonts. In other words, the abrasive occlusion, apart from the mechanisms generated by dysmorphoses, seems partly related to diet, and also to para-masticatory functions. In the case of a strong abrasion of the teeth, the distinction is *a priori* easy to make. This is not the case with abrasion on the anterior guide function, because to the naked eye no distinction is possible without the use of a fine tracing (MEB, for example). It therefore seems imperative to analyze not only the dental joints and their structural dynamics from the recent and contemporary history of the Maghreb and Sahara countries, but also the articulations of the fossil populations of anatomically modern humans.

16.5. Occlusal equilibrium and adaptation of regional morphotypes

The most effective method for understanding the mechanisms of craniofacial and occlusal growth and the rhythms of tooth eruption is the kinetic approach to the craniofacial puzzle and the antero-posterior and lateral movements of the dental joint. Several fossil and historical populations have followed this concept, starting with the setting up of a data bank analyzing archeological populations from medieval excavations in the south-east of Paris. For claiming to have a better understanding of occlusal dynamics or dental joints while misunderstanding its initial basic cranial dynamics is a challenge and the reverse is true.

16.5.1. *In the Paris Basin*

Funeral complexes from the Middle Ages in south-eastern Paris are estimated to contain nearly 3,000 skeletons of all ages. The analysis of several hundred skeletons among these groups serves us today as a reference since numerous studies have been carried out to answer the various mortality hypotheses (occlusion, craniofacial biodynamics, structure, posture, malformations, various skeletal diseases, alveolar diseases, epidemics, etc.). In addition, and in order to associate as many specialists as possible with this problem, we have set up a series of international meetings whose primary idea is transdisciplinarity (*International Symposium of Paleontology and Paleoanthropology, Créteil 1999, Symposium on Dental Rhythms, Créteil 2006, Osteopathy and Transdisciplinarity Symposium, École Supérieure d'Osteopathie de Paris, 2012* and *Frehops Symposium planned for 2021*). These meetings have not only shown the growing interest in the techniques and methods used, but above all have highlighted the lack of knowledge on the part of anthropologists of the analysis parameters and research tools used to highlight an anomaly, an asymmetry or a dento-skeletal dysmorphosis.

Among the 10 or so medieval and modern necropolises studied in an anthropological (morphotypes, craniofacial architecture, occlusion, postures, etc.) and health (skeletal and dental paleopathologies, epidemiology, stress states, developmental anomalies, etc.) framework, the five are among the most important (Hadjouis and Thillaud 1994; Hadjouis 1999, 2009, 2012c, 2012d, 2012e, 2017, 2018, 2019b; Hadjouis and Corbineau 2009; Hadjouis *et al.* 2009, 2015a, 2015b). These five are of interest with regard to the importance and frequency of extension and flexion dysmorphoses. These are the necropolis of Ivry-Parmentier 2, the medieval and modern churches of Chevilly-Larue, La Queue-en-Brie, Chennevières-sur-Marne and the Huguenot cemetery of the Charenton temple in Saint-Maurice. Among these five populations, three of them seem significant to us.

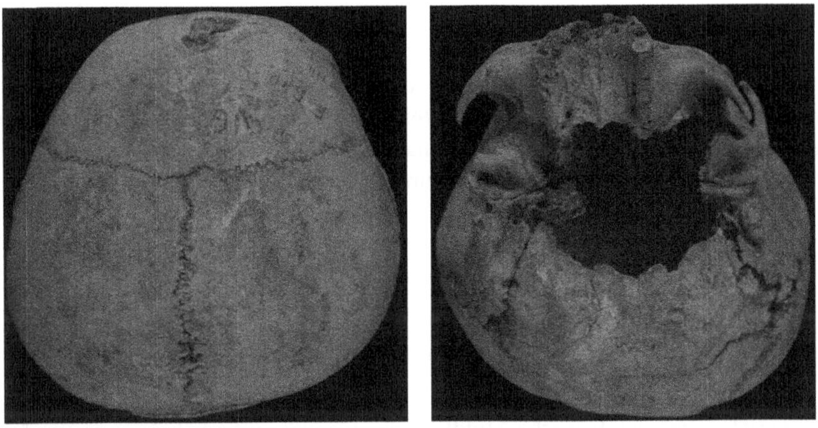

Figure 16.17. *Upper and lower views of a human skull from the medieval necropolis of La Queue-en-Brie, morphotype identical to that of the population of Chennevières-sur-Marne. Note the large width of the temporal walls and the antero-posterior reduction of the skull, due to the hyper-brachycephaly of this specimen and the excessive spheno-occipital flexion (© Hadjouis)*

The craniofacial morphotype represented in the medieval population of La Queue-en-Brie on the Brie Plateau (Figure 16.17) joins that of the neighboring villagers of Chennevières-sur-Marne (Figure 16.18). The anthropological results of the latter two populations suggest a kinship relationship, as their close morphologies seemed to be totally different from the morphotype of the central plain and western plateau sites of Longboyau. Two essential biometric and dynamic parameters of craniofacial and occlusal growth were distinguished in the population of La Queue-en-Brie: maximum skull width and a structural picture of excessive basic cranial flexion (Hadjouis 2017). These parameters reflected a greater brachycrania (37%) followed by hyperbrachycephalic (12%) and ultrabrachycephalic (8%) cranial forms, as well as the presence of class III dysmorphosis most often involving pro-mandibulia (too much forward projection of the mandible) (Cho 2005).

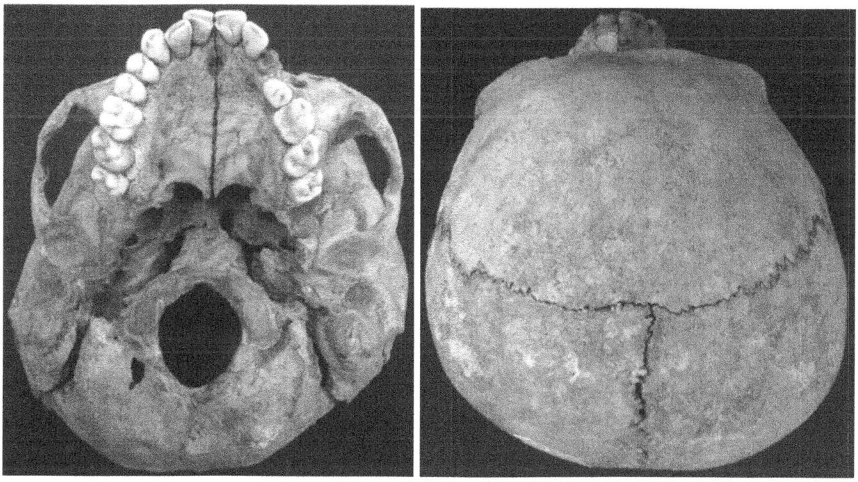

Figure 16.18. *Lower and upper views of a man's skull from the medieval and modern necropolis of Chennevières-sur-Marne. Note the great width of the temporal walls and the antero-posterior reduction of the skull, due to the hyper-brachycephaly of this specimen and the excessive spheno-occipital flexion (© Hadjouis)*

The morphometric characteristics of the Chevilly-Larue population point toward a mixed cranial morphology, but above all with mesocephalic and dolicocephalic tendencies, with class II type occlusion (cranial extension with slight mandibular retrognathia) (Figure 16.19) and generally medium to large size (Hadjouis 1998b, 1999; Hadjouis and Katz 1998). However, comparative craniofacial structural analyses suggest a strong predominance of extension-type dysmorphosis (21%) (Figures 16.21 and 16.22) over flexion-type dysmorphosis (6%) (Figures 16.20 and 16.23), whereas in the La Queue-en-Brie population, flexion far outweighed extension (49%), given the brachy and hyperbrachycephalic morphotypes of the latter (Cho and Hadjouis 2005). Craniofacial asymmetries were of the order of 29% (Hadjouis *et al.* 2015). Certain populations with smaller numbers, such as the necropolis of the Parvis Saint-Christophe in Créteil, suggest structural and occlusal trends of class II type (Hadjouis 1999; Ardouin *et al.* 2010).

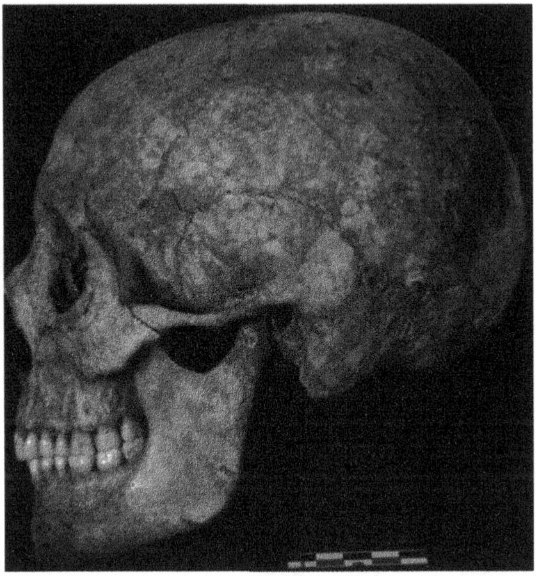

Figure 16.19. *Male skull in left lateral view from the necropolis of the church of Sainte-Colombe in Chevilly-Larue showing a psalidodental joint responding to a class II structural specimen (© Hadjouis)*

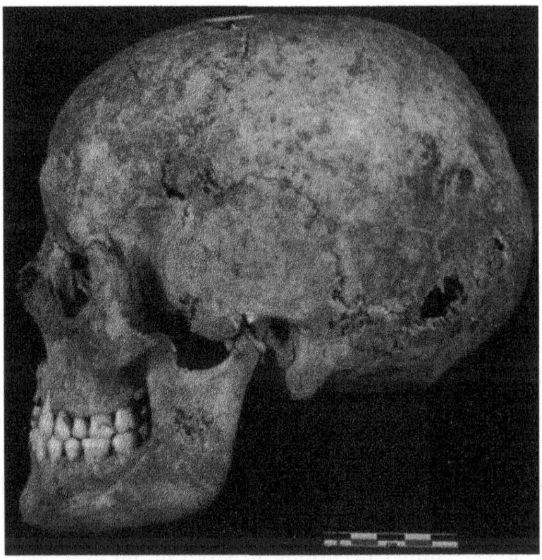

Figure 16.20. *Male skull in left lateral view from the necropolis of the church of Sainte-Colombe in Chevilly-Larue showing a labidodental dental articulation corresponding to a class III structural specimen (© Hadjouis)*

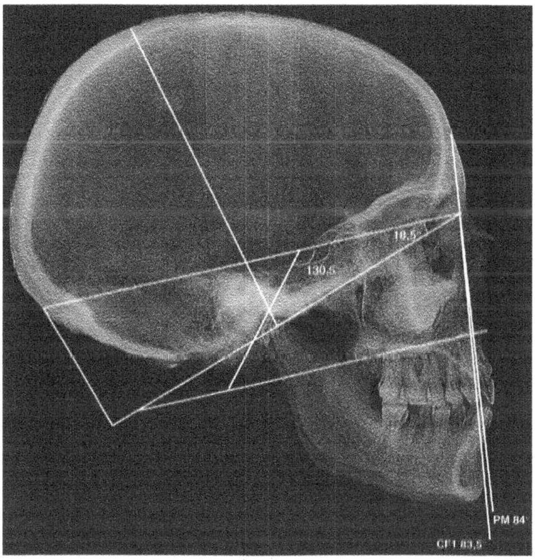

Figure 16.21. *Teleradiography of a profile of a female skull from the necropolis of the church of Sainte-Colombe in Chevilly-Larue, showing the weakness of the spheno-occipital tilt, the sphenoid angle is open at more than 130°, the angle of the face is closed at 18°, giving a class II structural specimen (© Hadjouis, Katz)*

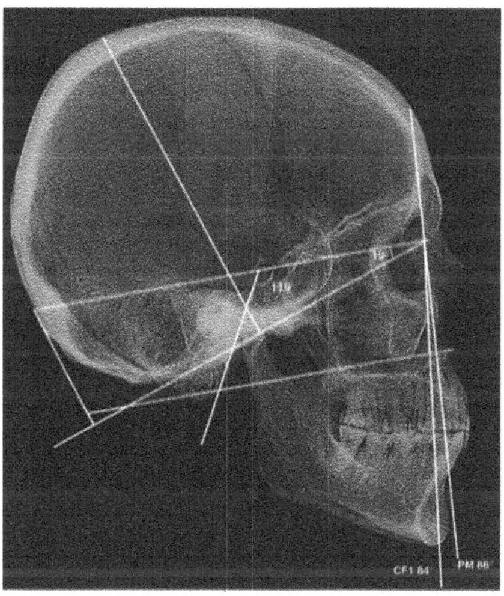

Figure 16.22. *Teleradiography of a profile of a female skull from the necropolis of the church of Sainte-Colombe in Chevilly-Larue. Despite a sphenoidal angle closed at 119°, an angle of the face open at 19°, the spheno-occipital tilt seems to be out of agreement with a face with a class II occlusion type with clearly an incisive end-to-end dental bite (© Hadjouis, Katz)*

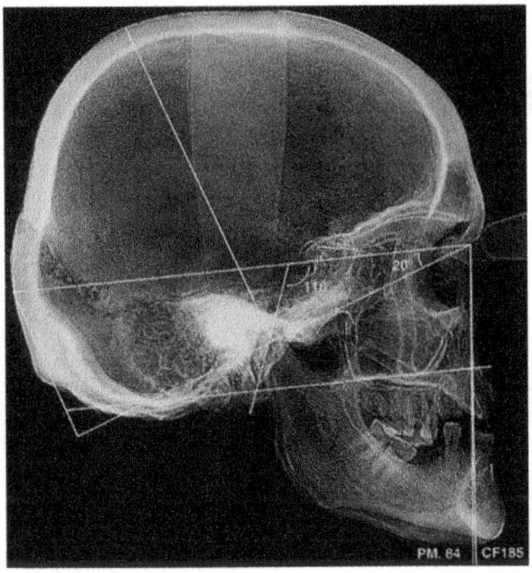

Figure 16.23. *Teleradiography of a profile of a man's skull from the necropolis of the church of Sainte-Colombe in Chevilly-Larue, showing a strong spheno-occipital flexion, a sphenoidal angle closed at 110°, a face angle open at 20°, the protrusive mandible is evident in this class III structural specimen (© Hadjouis, Katz)*

16.5.2. *In the Maghreb countries*

The craniofacial structural analysis and the occlusal posture of the anatomically modern humans of the Maghreb, in this case those of Afalou (around from – 16,000 to – 20,000 years), the Mechtoids and the Proto-Mediterraneans from the 8th millennium followed analytical parameters tested on medieval populations in the Paris Basin. In addition, a structural analysis by teleradiography was launched on a living Franco-Algerian population in order to measure the variability of the living morphotype within a wide geographical distribution, to compare it with other Mediterranean and African populations, and finally to compare it with the fossil populations of the same regions. Such an exercise required the assistance of partners from several countries (maxillofacial institutes in Paris, the University of La Sapienza in Rome, university hospitals and research centers in Algeria and the Universities of Abidjan and Dakar). This partnership is operational and is beginning to deliver its first results.

While the development of such techniques within certain teams is now mastered, this problem has highlighted shortcomings within therapeutic institutions whose concern is the care and esthetics of the face and its facial harmony. In other words, there are still today in the university establishments of Europe, orthodontic and dentofacial orthopedics training courses whose causes of dental or dento-skeletal imbalance in children are still sought in the oro-facial environment alone. Rare are those that relate to the base of the skull, the spine, the pelvis or the skeleton of the limbs. What can be said of such a practice in southern countries where only the major hereditary pathologies are confronted with real care, such as class III pathologies treated in particular in certain establishments in the Maghreb.

16.5.3. *Occlusal balance and the regional morphotype in the Maghreb and Sub-Saharan Africa*

Maghrebian and African orthodontic therapists quickly understood the interest of developing as quickly and efficiently as possible digital cephalometric analyses on their respective populations, given that in some countries one individual out of two requires orthodontic treatment. The French meetings that have been held for several years in Rabat, Casablanca, Sousse, Tlemcen, Algiers and Batna in collaboration with French therapists and those organized in Paris and elsewhere with their African colleagues, particularly in Abidjan and Dakar, have made significant progress not only in terms of the techniques used (analysis parameters, cephalometric methods, etc.) but also on the highlighting of structural specimens (class II and class III type dysmorphoses).

This preamble on maxillo-mandibular dysmorphoses (dento-dental or dento-skeletal) of North African and African populations clearly shows how well the discipline and its advanced techniques, like dentofacial orthopedics and dental surgery, are essential in emerging or developing countries, especially if the skull and face are studied according to structural dynamic parameters. However, it is useful to remember that the care given, based on a cephalometric model, is often carried out without the structural knowledge of the given population, I mean as an anthropological population. What are called by morphotype or structural type (Europoid morphotype or facies, Mediterranean morphotype or facies, Asian morphotype or facies, African morphotype or facies, etc.) are anthropological groups that have occupied geophysical spaces for thousands of years, even millions of years. Their structural trace found in the form of skeletons in necropolises reflects a paleoanthropological morphology of one or more structural types that have lived in these regions and which show the different evolutionary stages of the face and base of the skull and thus the prognathic or orthognathic alveolar character. In the Iberomaurusian populations of Algeria and Morocco, the spheno-occipital flexion by the closing of the sphenoid angle and the descent of the glenoid cavities favored the anterior projection of the mandible; however, despite the absence of anterior teeth of the maxilla caused by extraction, the dental articulation is in a logic of psalidodontia/labidodontia (Figures 16.24–16.26). This primordial characteristic of the craniofacial and occlusal morphology, which is nowadays found as a morphological type in the different geophysical regions, is sometimes even ignored by Western therapists.

The inclusion of comparative morphological and cephalometric data from north to south (Mediterranean regions of Northern Africa, Saharan regions, regions south of the Sahel) shows how striking the results of profile teleradiography are. Is there an occlusal difference in the profiles (orthonormal profile, non-pathological class I type) between the Mediterranean and Saharan Maghreb populations, and then between those of the Sahara, the Sahel and West Africa? Is there a craniofacial morphological similarity between African populations, and if so, which ones? What is the share of the mixing of populations resulting from migrations (south-north, north-south and east-west migrations)? Are there several morphotypes in North Africa, coming from ancient populations (prehistoric, protohistoric, ancient and medieval)?

All these questions have been debated for several years and a great deal of paleoanthropological and paleo-odontological work has been devoted to some of them. Many questions still remain unsolved due to a lack of numerical data, unpublished data, or simply by a certain lack of interest in these kinds of specialties.

The countries mentioned above that are beginning to modernize their equipment based on the new digital technologies should necessarily take into account the indigenous morphotypes that respond to a specific craniofacial structure. In this case, the dysmorphoses will reflect an abnormality in the structural morphotype of the said geographical region. It is obvious that certain occlusal shifts do not respond to a structural norm due to the ethnic and socio-cultural behaviors of certain localities (endogamy, genetic defects, tooth extraction, etc.).

If we refer to the contemporary Europoid populations of Ile-de-France, the analyses carried out on more than 500 teleradiographic photographs already show the structural tendencies and dysmorphoses encountered in particular by a quasi-generalized trend in class II (Carré *et al.* 2006).

The work of Chamla (1970) on the Arab-Berber populations of North Africa showed, on the basis of certain cranial indices, notably the cephalofacial index, that about 80% of these populations are Mediterranean, with a representation of three sub-types:

– an Atlanto-Mediterranean type (long to very long mesocephalic face);

– an Iberian-insular type (dolicocephaly with medium to long face);

– a Saharan type (dolicocephaly, with long to very long face).

There are also two anthropological minorities:

– brachycephalus with elongated face;

– Meso to short-faced dolichocephalous heads reminiscent of the Mechta-Afalou, the ancient Cromagnoids of the Upper Paleolithic, found in Algeria and Morocco.

The first results of the structural analysis by teleradiography of the Maghreb profile show an increase in the base of the skull often in class II individuals (in retromandibulia), a strong hyperdivergent tendency (long face) as well as a maxillary structure which is always orthognathic (Hadjouis 2011). The comparisons between the leukoderms of the North and the Melano-African populations allow us to appreciate the morphological similarities or certain craniofacial differences that may exist in the facial structures and to identify the dysmorphic tendencies.

Thus, the Saharan populations of Algeria and Morocco often demonstrate dolichocephaly with long to very long faces, sharing the same occlusal profile, i.e. an elongated base in normocclusion or with mandibular retrusion and similar orthognathy. In Chad (Sahelian country), a mesodivergent tendency (middle face) can be found, with a maxillary structure showing moderate alveolar prognathism accompanied by bi-alveolar protrusion and sometimes a bi-alveolar protrusion and bi-procheilia (Beugre *et al.* 2005; Sonan *et al.* 2006). The Ivorian population presents a promaxillium in normocclusion or class II and shares with the Senegalese a significant alveolar prognathism coupled with bi-procheilia (Beugre *et al.* 2005).

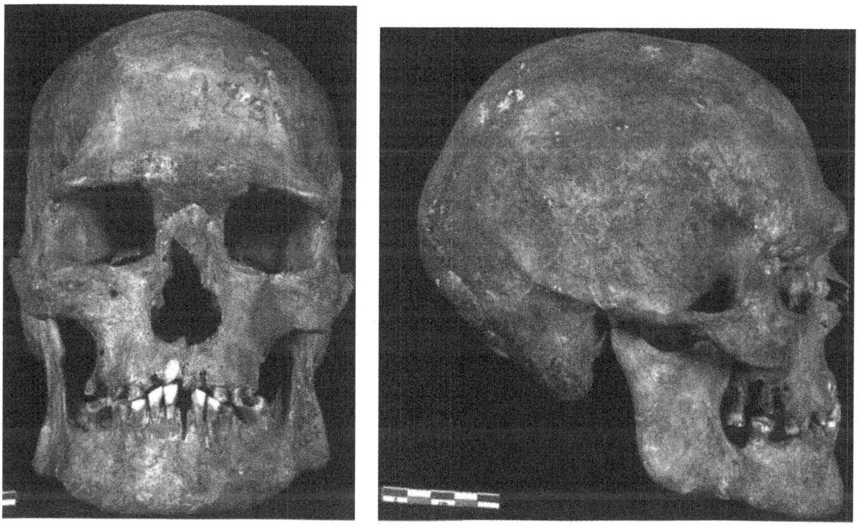

Figure 16.24. *Facial and right lateral views of a skull of a robust Iberomaurusian man from Afalou showing the failed extractions of the upper incisors and the inflammation of the latter and carious disease of the jugal teeth (© Hadjouis)*

All in all, it appears that while occlusal shifts are now identified and recognized in the Maghreb and African countries according to appropriate analysis parameters, it is nevertheless true that cephalometric diagnoses are often made without taking into account the craniofacial structural morphotype (dolicocephaly, mesocephaly, brachycephaly, hyperbrachycephaly, etc.) where the three levels of the face (lower, middle and upper) coordinated at the base of the skull take on their full meaning for better functional therapy. In addition, the new digital cephalometric instruments used in Western countries have taken such a place that the basic clinical examination is sometimes relegated to the background (Philippe 2008).

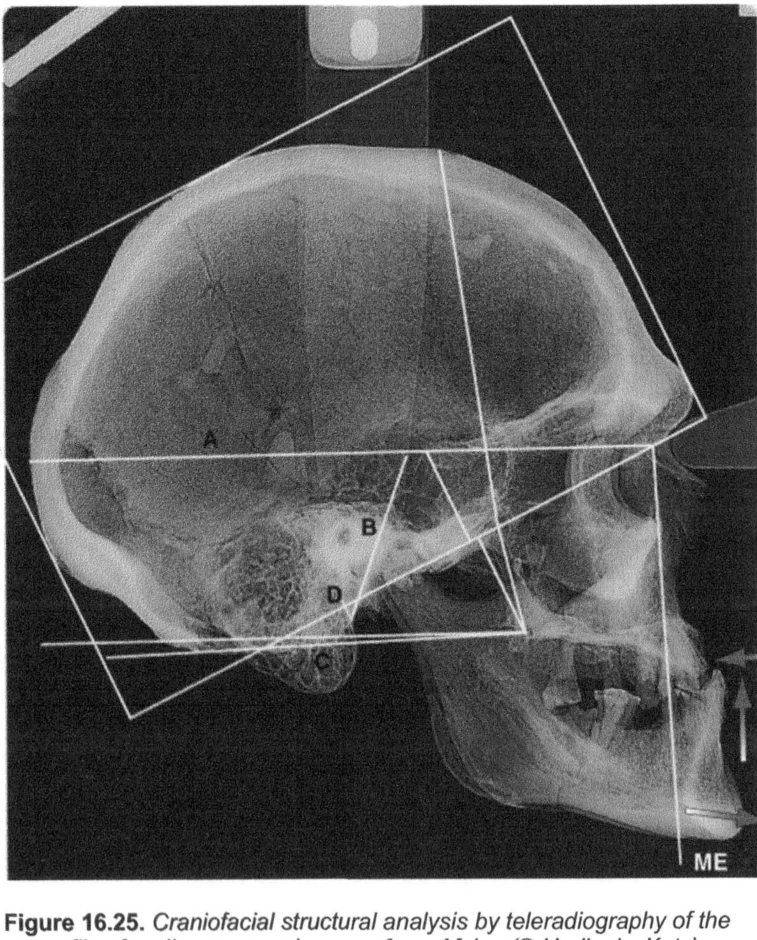

Figure 16.25. *Craniofacial structural analysis by teleradiography of the profile of an Iberomaurusian man from Afalou (© Hadjouis, Katz)*

COMMENT ON FIGURE 16.25.– *Analysis of the photograph shows spheno-occipital flexion by closing of the sphenoid angle and the descent of the glenoid cavities promoting the anterior projection of the mandible. The extraction of the upper incisors promoted the thrust in height of the lower teeth; however, despite the absence of teeth, the dental joint seems to be labiodental/psalidodental.*

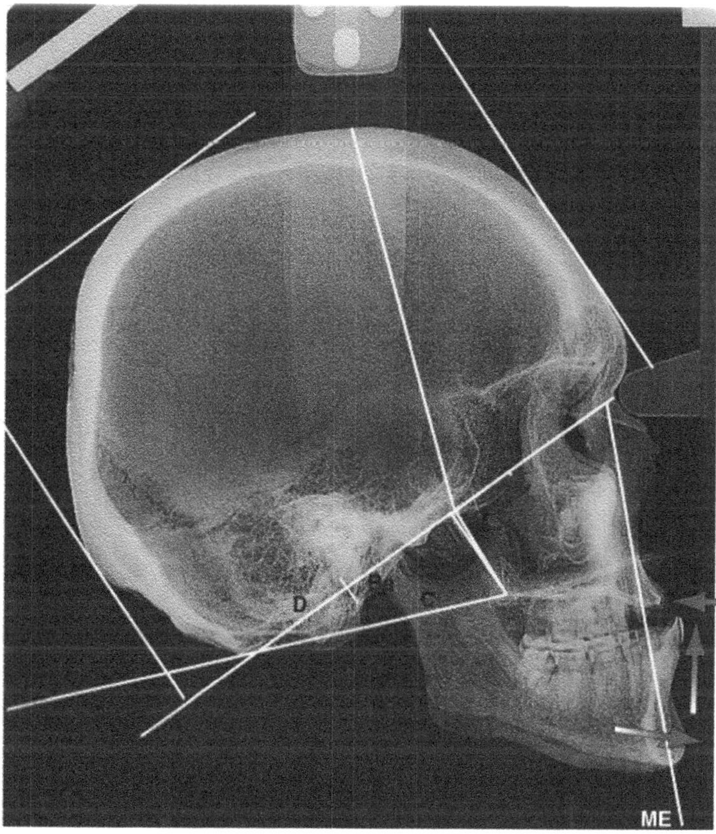

Figure 16.26. *Craniofacial structural analysis by teleradiography of the profile of an Iberomaurusian woman from Afalou (© Hadjouis, Katz). For a color version of the figure, seewww.iste.co.uk/hadjouis/skull.zip*

COMMENT ON FIGURE 16.26. – *Analysis of the photograph shows spheno-occipital flexion by the closing of the sphenoid angle and the descent of the glenoid cavities promoting the anterior projection of the mandible. The extraction of the upper incisors promoted the thrust in height of the lower teeth; however, despite the absence of teeth, the dental joint seems to be labidodental/psalidodental.*

17

Ignored Pathologies

17.1. Extremely rare craniofacial pathologies

The medical literature, particularly that devoted to children's diseases, reports a large number of craniofacial syndromes and dysostoses of genetic, hereditary or metabolic origin that also associate various disorders (mental disability, eye and dental disorders, aplasia of certain bone elements, etc.). In addition to these conditions, there is a mosaic of malformations followed by craniofacial puzzle disorders and a significant maxillo-mandibular imbalance. Some of these syndromes are presented here, described in regard to specimens found in excavations.

17.1.1. *Crouzon syndrome*

The paleopathologist helped by his experimentation in the interpretation of medical images was able to identify a certain number of them. Archeological excavations have yielded a few specimens bearing the symptoms of craniofacial dysostosis or Crouzon's syndrome, of autosomal dominant transmission, characterized by: an increase in cerebral volume due to premature closure of the coronal and sagittal sutures, hypoplasia of the maxilla, ogival deformation of the arch of the palate and facial prognathism. It should be noted that the craniofacial characteristics of this genetic disease can be confused with a craniostenosis-type malformation.

17.1.2. *Marfan syndrome*

During a preventive excavation against the wall of the military fort of Kremlin-Bicêtre (Val-de-Marne) three bodies of soldiers from the Franco-Prussian War of 1870 were unearthed with their military clothing. Among the three skeletons, a young individual in his early twenties displayed the characteristics of Marfan's syndrome. In terms of the face, a significant asymmetry of the type rotation lateral flexion toward the left-hand side affected the three levels of the face, resulting in a right unilateral mandibular dislocation and infracranial effects, notably an asymmetry of the upper cervical spine. All the features of this disease suggested this syndrome: facial and basic cranial asymmetry, mandibular dislocation, asymmetry of the cervical spine, medial fissure of the anterior arch of C1, carina thorax (*pectus carinatum*), scaphoid scapula, sacralization of L5 and *hallux valgus* (Hadjouis *et al.* 2020). It should be remembered that skeletons found in funeral contexts or in archeological habitats showing these symptoms are exceptional, because among the advantages of its identification, only excellent bone conservation allows us to arrive at a satisfactory diagnosis.

17.1.3. *Cranial thickening and Albers-Schönberg's disease*

The thickening of the skull and face, accompanied by dental eruption disorders, cannot be confusing if the analyses are detailed and meaningful, such as vitamin-resistant rickets, which shows porous patches (porous hyperostosis syndrome) or sometimes complete hyperossification of the skull and skeleton of the limbs (Hadjouis 2018). Teeth displaying rickets are another aid in the identification of the disease.

A case of exaggerated thickening of the skull and mandible, followed by a disorder in the rhythm of tooth eruption (numerous impacted teeth and poor tooth implantation), suggests the characteristics of a case of osteopetrosis (Albers-Schönberg's disease): hyperdense radiographic imaging, excessive mineral overload, mandibular disease, nasal swelling, bony osteomas, etc. (Hadjouis 2018). Like the previous syndrome, the retrospective diagnosis of osteopetrosis is practically unknown in the archeological literature.

17.1.4. *Torticollis*

Torticollis is a muscular disorder that affects the cervical muscles, particularly the unilateral retraction of the sternocleidomastoid muscle, making it impossible for the neck to turn. The 10:10 a.m. time sign (head tilted to one side, eyes looking in the other) is the medical jargon used to identify the shape of the baby's skull during childbirth. Whether it is of unknown pathogenesis or cervicalalgia, the muscular torticollis can become a bony torticollis. Among the anthropological collections in the Val-de-Marne is a skull from the medieval necropolis in La Queue-en-Brie (Hadjouis 2017, 2018). The base of the skull is characterized by basic cranial asymmetry (retreat of the right hemi-cranium, exaggerated enlargement of the jugular fossa on the same side). But the most important feature is the narrowing of the *foramen magnum*, whose right occipital joint overhangs the spinal cord orifice. Let us not forget to mention that this condition, initially congenital, has had supra-cervical consequences, which we have just seen above, and infracranial irreversible unilateral asymmetries, affecting the entire cervical spine and clavicle.

17.1.5. *Parietal thinning*

The unilateral or bilateral thinning of the parietals (*depressio biparietalis circumscripta*) is described following Manouvrier on skulls whose depression of the external specimen evokes a traumatic lesion. It is useful to remember that T-shaped or sincipito-obeliatic sincipital dysplasia is not easy to identify, since it is confused with the practice of trepanning. However, its prevalence in archeological sites directs the population toward an identification of parental relationship, with dysplasia being re-evaluated as an epigenetic trait (Lacombe 2004).

In the necropolis of Chennevières-sur-Marne, there is a fragmentary skull belonging to an adult male. The various fragments that have been reassembled come from the vault (parietal, frontal and occipital). The pieces preserved without being reassembled come from the two temporals and the basi-sphenoid. The bones of the face, maxilla and mandible are absent. It was the presence of a fracture of the vault on one of the fragmentary pieces that necessitated its reassembly. Indeed, the right parietal retained a double fracture of the depressed type. The successful welding of what appear to be the two injuries was not fatal to this individual; the first injury was perpendicular to the sagittal axis of the cranial vault and measured 48 mm long and 8 mm wide in the middle. Its depth was about 1 mm and the sinking seems significant in its middle, where the fracture widened. The second fracture, of the same type, crosses the first in its last third and is oriented from front to back. The absence of part of the parietal in its anterior zone deprives us of the extent in length of the fracture. However, the depression appears to be much deeper and wider (17 mm). The fragmentary parietal part reveals a bony slice where the outer type and the diploë are clearly well crushed (Figure 17.1).

The retrospective diagnosis of this double fracture remains, however, marred by uncertainty, and the differential diagnosis (which could even be the retrospective diagnosis) leans seriously toward an anatomo-pathological etiology, in this case sincipito-obelia dysplasia (Le Double 1903; Manouvrier 1903). Both dysplastic forms are part of parietal depressions of phenotypic expression. The one that comes close to our specimen given its obeliac localization is the second form which, as mentioned above, is of interest in this thinner parietal zone, probably corresponding to an epigenetic character (Lacombe 2004).

Figure 17.1. *Double angular depression on the parietal wall of a probably male adult skull. The diagnosis of this atrophy is part of parietal thinning and leans toward phenotypic expression obeliac dysplasia (© Hadjouis)*

The medieval and modern necropolis of the Saint-Pierre church in Chennevières-sur-Marne has yielded more than a hundred immature and adult individuals whose several cranial, dental and post-cranial characters suggest family relationships (sincipito-obelia dysplasia, anomalies in the rate of dental growth, Carabelli's tubercle, Bolk's tubercle, *cubitus valgus*, *genu varum*, chronic osteomyelitis, etc.) (Ardouin and Hadjouis 2014; Hadjouis 2020b).

17.1.6. *Scurvy*

Scurvy is a hypovitaminosis C, first identified in Scandinavian sailors in the 18th century. At the oral level, it manifests itself by gingival hemorrhages and periodontitis, whereas at the skeletal level, the bones of immature individuals are weakened by a form of osteoporosis.

Like the previous diseases, this extremely rare condition is present in the neocropolis of the Saint-Pierre church in Chennevières-sur-Marne on a dozen children's skeletons buried in the modern levels of the cemetery (18th–19th centuries).

The temporary and permanent maxillo-mandibular teeth show brownish or orange pigmentation due to infection from gingivitis hemorrhages (Figure 17.2). The bones of the limbs of the same individuals were affected by a form of osteoporosis, coupled with osteoperiostitis. Some individuals showed symptoms of chronic childhood osteomyelitis (Ardouin and Hadjouis 2014; Hadjouis 2020b).

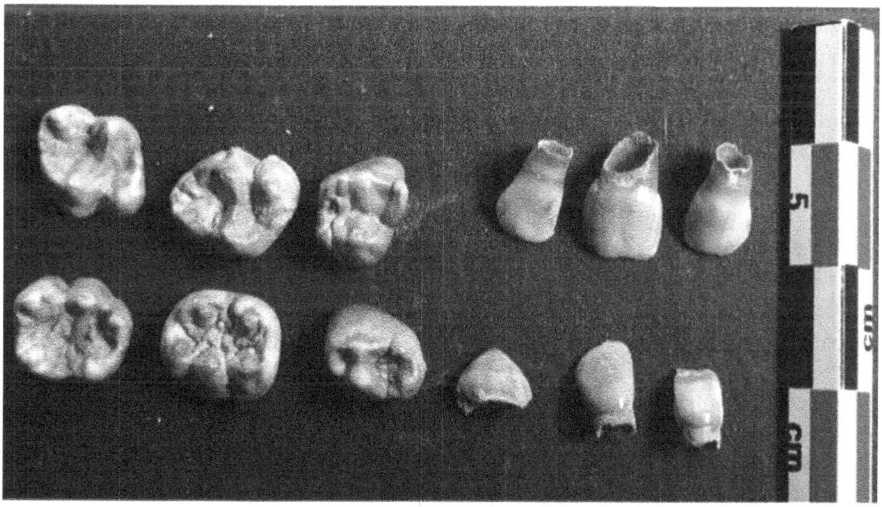

Figure 17.2. *Upper-inferior incisivo-canine and milky premolar series and M1 buds of a child aged about 12 months from the modern levels of the necropolis of Chennevières-sur-Marne. The brown and brownish pigmentation of the dental crowns in relation to lesions of the skeleton of the limbs (juvenile osteoporosis, childhood osteomyelitis) suggests a diagnosis of scurvy by gingival hemorrhage (© Hadjouis)*

17.2. The oldest therapeutic practice: trepanning

The work of the military physicians of colonial Algeria in the 19th century published on the therapeutic practices, curious to say the least, but skillfully orchestrated on the populations of the Aurès mountains and Kabylia, discussed the heads of villagers suffering from headaches or other head-related pain.

Among the works of military physicians, the accounts of Dr. Martin (1867) "On the trepanning of the skull as practiced by the kabyles of the Aurès" and those of Dr. Paris (1867) "On the cephalic trepanning practiced by the physicians of the Aouress (Aurès)" described both the therapeutic act itself and the instruments used. Other authors, such as Dr. Malbot, medical officer of the military hospitals of Algeria, and Dr. Verneau, anthropologist at the Institute of Human Paleontology in Paris, describe in detail the trepanning practice of these villagers who never published a treatise on human anatomy. The tools used for such acts consist of an auger, a saw and a few other smaller pieces. The auger or *brima* is a steel rod, fixed in a wooden handle, made of two sets that fit together. The saw or *menchar* is made of steel, 15 cm long, with a bent tip that ends in saw teeth. The other accessory instruments were a knife, razor, single or double hook and palette knife. The texts mention that morbid symptoms such as headaches, convulsions and paralysis caused by the presence of a bone made ill by a contusion or a fracture were well identified by the trepanner and the diseased bone therefore needed to be removed at all costs (Malbot and Verneau 1897). Among the forms described, square trepanning was the most common, in other words drilling with a saw in one or two sessions in the case of major skull trauma.

We had the opportunity to study one of the specimens trepanned by the Chaouia people of the Aurès, preserved in the collections of the Musée de l'Homme in Paris. The numerous perforations of the parietal that preceded the square trepanning were still intact and show indisputably both the historical reality of such a surgical act and the care given to such practices, which have been used since prehistoric times.

Cranial trepanation has been known since the Upper Paleolithic of the Maghreb (rock shelters of Taforalt in Morocco and Afalou Bou Rhummel in Algeria). Sites dating from the Neolithic and more recent sites yielding skulls that have undergone empirical surgery numbers around 500 specimens (Jamet 2000), but some have been confused with parietal thinning (*depressio biparietalis circumscripta*) or sincipito-obelial dysplasia corresponding to epigenetic features (Lacombe 2004).

Dr. Broca's work on cranial trepanning for magical-therapeutic purposes in relation to convulsions was accepted during his lifetime, but was partially rejected after his death in 1880. And it is the synthesis of Lucas-Championnière's work (1912), an advocate of decompressive trepanning, that was one of Broca's main detractors (Charon and Thillaud 2010).

In his long research on empirical skull trepanning, Dastugue distinguished between shocking lesions opening the endocranial cavity to the outside, which are true *in vivo* surgical procedures, and non-perforating lesions (incomplete trepanning, rather called cutting) whose depth never opens the endocranium (Dastugue and Gervais 1992).

Conclusion

In recent years, we have brought back to the forefront concepts that are so often ignored, sometimes misunderstood, such as the holistic concept in anatomy and the interdisciplinarity of the biological sciences. While the former is committed to structural dento-skeletal treatment in its postural globality, the latter makes itself useful and indispensable to the understanding of complex phenomena. Is it not risky to resort to an analysis of craniofacial and post-cranial morphogenesis of only adult bone and/or dental elements, neglecting or ignoring ontogenic growth and its stages, sometimes accelerated, sometimes slowed down? Examples of this type are numerous and also concern the notions of intra-specific variation, or bone growth in children without taking into account metabolic, hereditary and familial or congenital damage. And yet, if the History of Science were taught on a regular basis within the natural sciences, we would not be surprised that these notions have been lavished since Geoffroy-Saint-Hilaire, then Georges Cuvier and their followers, who gave composed anatomy its letters of nobility. Indeed, it is thanks to the comparative anatomy of Vertebrates that such morphogenesis, functions and other mechanisms of membranes and chondroses are explained. And we have never tired of replacing the teachings of Dr. Le Double on the importance of bone variations from a zoological point of view and those of Dr. Dastugue on the boundaries of the normal, the abnormal and the pathological. Dr. Cazeau, an orthopedic surgeon, will in turn say that the normal and the useful should not be ignored either (Cazeau 2020).

The launch of the "Comparative Anatomy and Posture of Animals and Humans" set at ISTE Ltd is not by chance. This book of the set, which is devoted to the skull of quadrupeds and bipeds, shows how essential the notions of comparative anatomy and posture are for the understanding of the evolutionary mechanisms of hominids, especially during periods when some were arboreal and terrestrial, others permanent terrestrial.

This book is a summary of 40 years of research on the study of the quaternary fauna of Africa, the Mediterranean Basin and Europe, and on the anatomically modern humans of the Maghreb and the Paris Basin, as well as their state of health. The latter state has shown the importance that an infection, inflammation, trauma or genetic damage can have on an isolated bone or on the whole skeleton.

With this set at ISTE Ltd, certain terms used in anatomy, paleontology and paleoanthropology such as posture, balance, gravitation, asymmetry, spheno-occipital flexion, trajectory and growth will have a particular resonance. And it is not for nothing that the comparative anatomy of the quadruped species, in this case the Ungulates, and the human species in the standing position are put in parallel, because the former will have mostly, if not totally, balanced centers of gravity and a cantilever head, unaware of the asymmetries and craniofacial dysmorphoses, while the others will have an unbalanced posture, with innumerable asymmetries and malocclusions once the vertical position is acquired.

References

Abbazzi, L. (2004). Remarks on the validity of the generic name *Praemegaceros* Portis 1920, and an overview on *Praemegaceros* species in Italy. *Rend. Fis. Acc. Lincei*, 9(15), 115–132.

Abbazzi, L. and Masini, F. (1997). *Megaceroides solihacus* and other deer from the middle Pleistocene site of Isernia La pineta (Molise, Italy). *Bolletino della Societa paleontologica Italiana*, 35(2), 213–227.

Alexeev, V.P. (1978). *Paleoantropologia zemnogo shara i formirovanie chelove-cheskikh ras.* Nauka, Moscow.

Altiéri, M. (1984). *Biomécanique et pathologie crânienne en étiopathie.* Ethio-sciences, Geneva.

Ambrosetti, P. (1967). Cromerian fauna of the Rome area. *Quaternaria*, 9, 267–284.

Angle, E.H. (1899). Classification of malocclusion. *Dental Cosmos*, 248–264/350–357.

Anthony, J. (1973). *Cours d'Anatomie dentaire comparée.* Éditions CdP, Paris.

Aoudia-Chouakri, L. (2003). Analyse architecturale par téléradiographie des crânes de la grotte de Taforalt (Pléistocène supérieur, Maroc). Comparaison avec les hommes d'Afalou Bou Rhummel (Paléolithique supérieur, Algérie). DEA, Muséum national d'Histoire naturelle, Paris.

Aouraghe, H. (2001). Contribution à la connaissance des faunes du Pléistocène supérieur du Maroc : les Vertébrés d'El Harhoura (Témara), comparés à ceux de plusieurs sites du Maghreb. Paléontologie, Archéozoologie, Paléo-environnement, Biochronologie. PhD Thesis, Université Mohamed Ier, Oujda.

Aouraghe, H. and Débénath, D. (1999). Les Équidés du Pléistocène supérieur de la grotte Zouhrah à El Harhoura, Morocco. *Quaternaire*, 10(4), 283–292.

Arambourg, C. (1931). Observations sur une grotte à ossements des environs d'Alger. *Bulletin de la Société d'histoire naturelle d'Afrique du Nord*, XXII, 169–170.

Arambourg, C. (1932). Note préliminaire sur une nouvelle grotte à ossements des environs d'Alger. *Bulletin de la Société d'histoire naturelle d'Afrique du Nord*, 23(7), 154–162.

Arambourg, C. (1938). La faune fossile de l'Aïn Tit Mellil (Maroc). *Bulletin de la Société préhistorique du Maroc*, 1–4.

Arambourg, C. (1947). *Mission Scientifique de l'Omo : contribution à l'étude géologique et paléontologique du Bassin du lac Rodolphe et de la Basse Vallée de l'Omo. Deuxième partie : paléontologie*, volume I, instalment 3, 231–562. Éditions du Muséum, Paris.

Arambourg, C. (1954). L'Hominien fossile de Ternifine (Algérie). *Académie des Sciences*, Paris, 239, 893–895.

Arambourg, C. (1955). Une découverte récente en Paléontologie humaine, l'*Atlanthropus* de Ternifine (Algérie). *Quaternaria*, 2, 5–13.

Arambourg, C. (1979). *Vertébrés villafranchiens d'Afrique du Nord (Artiodactyles, Carnivores, Primates, Reptiles, Oiseaux)*, volume 2. Édition de la Fondation Singer-Polignac, Paris.

Arambourg, C. and Coppens, Y. (1967). Sur la découverte dans le Pléistocène inférieur de la Vallée de l'Omo (Éthiopie) d'une mandibule d'Australopithécien. *Comptes rendus de l'Académie des Sciences*, 265, D, 589–590.

Arambourg, C. and Hoffstetter, R. (1954). *Découverte en Afrique du Nord, de restes humains du Paléolithique inférieur*. Académie des Sciences, 73–74, Paris.

Arambourg, C. and Hoffstetter, R. (1955). *Le gisement de Ternifine. Résultats des fouilles de 1955 et découvertes de nouveaux restes d'Atlanthropus*. Académie des Sciences, 431–433, Paris.

Arambourg, C. and Hoffstetter, R. (1963). *Le gisement de Ternifine*. Masson, Paris.

Arambourg, C., Boule, M., Vallois, H., Verneau, R. (1934). *Les grottes paléolithiques des Beni Segoual (Algérie)*. Institut de Paléontologie humaine, Paris.

Ardouin, S. and Hadjouis, D. (2014). Chennevières-sur-Marne : église Saint-Pierre, 55, rue du général de Gaulle. Report, Conseil général du Val-de-Marne, Créteil.

Ardouin, S., Hadjouis, D., Arroyo-Bishop, M. (2009). Étude archéologique de la Seine dans le Val-de-Marne du Paléolithique au haut Moyen Âge. Study, Service Archéologie, Val-de-Marne.

Ardouin, S., Galmiche, T., Hadjouis, D., Fleury, L. (2010). Extension et chronologie de la nécropole médiévale de Créteil (Val-de-Marne). *Revue archéologique d'Île-de-France*, 3, 149–176.

Argant, A. (1991). Carnivores quaternaires de Bourgogne. *Doc. Lab. Géol.*, 115.

Argant, A. (1996a). Sous-Famille des *Ursinae*. In *Les Grands Mammifères plio-pléistocènes d'Europe*, Guérin, C., Patou-Mathis, M. (eds). Masson, Paris.

Argant, A. (1996b). Sous-Famille des *Felinae*. In *Les Grands Mammifères plio-pléistocènes d'Europe*, Guérin, C., Patou-Mathis, M. (eds). Masson, Paris.

Arsuaga, J.-L., Martinez, I., Lorenzo, C., Gracia, A., Munoz, A., Alonso, O., Gallego, J. (1999). The human cranial remains from Gran Dolina Lower Pleistocene site (Sierra de Atapuerca, Spain). *J. Hum. Evol.*, 37, 421–457.

Ascenzi, A., Biddutu, I., Cassoli, P.F., Segre, A.G., Segre-Naldini, E. (1996). A calvarium of late *Homo erectus* from Ceprano, Italy. *J. Hum. Evol.*, 31, 409–423.

Asfaw, B., White, T., Lovejoy, O., Latimer, B., Simpson, S., Suwa, G. (1999). *Australopithecus garhi*: A new species of Early Hominid from Ethiopia. *Science*, 284(5414), 629–635.

Audoin-Rouzeau, F. (1994). Bêtes médiévales et familiarité : animaux familiers de l'esprit, animaux familiers de la vie. *Anthropozool.*, 20, 11–40.

Azzaroli, A. (1953). La systematica dei Cervi giganti e Cervi nani delle isole. *Atti. Soc. Toscana Sc. Nat.*, 59.

Azzaroli, A. (1979). Critical remarks on some giant deer (genus *Megaceros* Owen) from the pleistocene of Europe. *Paleontolog.*, 5–16.

Azzaroli, A. and Mazza, P. (1992). On the possible origin of the Giant Deer genus *Megaceroides*. *Rend. Fis. Accad. Liencei.*, 9(3), 23–32.

Bagtache, B. and Hadjouis, D. (1983). Deux nouvelles espèces d'*Equus* (*Mammalia, Perrissodactyla*) dans le gisement atérien des Phacochères (Alger, Algérie). *Lybica*, 30–31, 165–186.

Bagtache, B., Hadjouis, D., Eisenmann, V. (1984). Présence d'un *Equus* caballin (*Equus algericus n. sp.*) et d'une autre espèce nouvelle d'*Equus* (*Equus melkiensis n. sp.*) dans l'Atérien des Allobroges (Alger). *Comptes rendus de l'Académie des Sciences*, 298(II, 14), 609–612.

Ballesio, R. (1975). Étude de *Panthera (Leo) spelaea* (Goldfuss) nov. sb. sp. (*Mammalia, Carnivora, Felidae*), du gisement du Pléistocène moyen des Abimes de la Fage à Noailles (Corrèze). *Nouv. Archiv. Mus. Hist. Nat.*, 13, 47–55.

Barone, R. (1976). *Anatomie comparée des mammifères domestiques : ostéologie*, volume 1. Vigot, Paris.

Bate, D.M.A. (1949). A new African fossil long-horned Buffalo. *Ann. Mag. Nat. Hist.*, 2(12), 396–398.

Bate, D.M.A. (1951). The mammals from Singa and Abu Hugar. *Fossils Mammals of Africa*, 2.

Bermudez De Castro, J.-M., Arsuaga, J.-L., Carbonell, E., Rosas, A., Martinez, I., Mosquera, M. (1997). A Hominid from the Lower Pleistocene of Atapuerca, Spain: Possible ancestor to Neanderthals and Modern Humans. *Science*, 276, 1392–1395.

Beugre, J.B., Beugre-Kouassi, A.M.L., Sonan, N.K., N'dindin-Guinan, B., Fareou, P., Faure, J., Djaha, K. (2005). Cépahalométrie comparative et appréciation esthétique faciale en Afrique noire subsaharienne. *Biom. Hum. et Anthropol.*, 23(1–2), 63–72.

Biraben, J.N. (1975). *Les hommes et la peste en France et dans les pays européens et méditerranéens : la peste dans l'Histoire*, volume 1. EHESS-CRH, Paris.

Biraben, J.N. (1976). *Les hommes et la peste en France et dans les pays européens et méditerranéens : les hommes face à la peste*, volume 2. EHESS-CRH, Paris.

Biraben, J.N. (1995). Les maladies en Europe : équilibres et ruptures de la pathocénose. In *Histoire de la pensée médicale en Occident : Antiquité et Moyen Âge*, volume 1, Grmek, M.D. (ed.). Le Seuil, Paris.

Björk, A. and Skieller, V. (1972). Facial development and tooth eruption. An implant study at the age of puberty. *Am. J. Orthodont.*, 62, 339–383.

Blanc, H.L. (1983). *Guide du cheval et du poney*. Delachaux et Nestlé, Lonay.

Bocquet, J.P. and Masset, C. (1977). Estimateurs en paléodémographie. *L'Homme*, 17(4), 65–90.

Boessneck, J., Jéquier, J.P., Stampfli, H.R. (1963). Seeberg Bürgachisse-Sud, 3ème partie. *Die Tierreste*, 117–196.

Bouchud, J. (1975). *La faune de Medjez II. Un gisement capsien de faciès sétifien, Medjez II, El Eulma (Algérie)*, 377–391, CNRS, Paris.

Bourguignat, J.R. (1867). *Sur un Ursus nouveau découvert dans la grande caverne du Thaya*. Bouchard-Huzard, Paris.

Bourguignat, J.R. (1868). *Notice prodromique sur quelques Ursidae d'Algérie*. Bouchard-Huzard, Paris.

Braga, J. (2005). Avant le genre *Homo*. In *Origine et évolution des populations humaines*, Dutour, O., Hublin, J.-J., Vandermeersch, B. (eds). Comité des travaux historiques et scientifiques, Paris.

Brunet, M., Beauvilain, A., Coppens, Y., Heintz, E., Moutaye, A.H.E., Pilbeam, D. (1995). The first *Australopithecine*, 2,500 kilometres west of the Rift Valley (Chad). *Nature*, 378, 273–275.

Brunet, M., Fronty, P., Sapanet, M., Viriot, L. (2001). Les hypoplasies dentaires chez *Australopithecus bahrelghazali*, hominidé pliocène du Tchad. In *La paléodontologie, analyse et méthodes d'études*, Hadjouis, D., Mafart, B. (eds). Artcom', Paris.

Brunet, M., Guy, F., Pilbeam, D., Mackaye, H.T., Likius, A., Ahounta, D., Beauvilain, A., Blondel, C., Bocherens, H., Boisserie, J.-R., De Bonis, L., Coppens, Y., Dejax, J., Denys, C., Duringer, P., Eisenmann, V., Fanone, G., Fronty, P., Geraads, D., Lehmann, T., Lihoreau, F., Louchart, A., Mahamat, A., Merceron, G., Mouchelin, G., Otero, O., Campomanes, P.P., De Leon, M.P., Rage, J.-C., Sapanet, M., Schuster, M., Sudre, J., Tassy, P., Valentin, X., Vignaud, P., Viriot, L., Zazzo A., and Zollikofer C., (2002). A new Hominid from the Upper Miocene of Chad, Central Africa. *Nature*, 418, 145–151.

Carbonell, E., Bermudez De Castro, J.-M., Arsuaga, J.-L., Diez, J.-C., Rosas, A., Cuenca-Bescos, G., Salar, R., Mosquera, M., Rodriguez, X.P. (1995). Lower Pleistocene Hominids and artefacts from Atapuerca-TD6 (Spain). *Science*, 269, 826–832.

Carré, H., Hadjouis, D., Katz, P. (2006). Croissance cranio-faciale des populations contemporaines d'Île-de-France et comparaison avec les populations médiévales du Val-de-Marne. Poster, Colloque Rythmes de croissance et fonctions manducatrices chez les hommes fossiles et modernes. Créteil, March.

Cazeau, C. (2020). *Foot Surgery Viewed Through the Prism of Comparative Anatomy*. ISTE Ltd, London, and John Wiley & Sons, New York.

Chaid-Saoudi, Y. (1984). Étude systématique du genre *Equus* (*Mammalia, Perissodactyla*) de l'Épipaléolithique de Columnata (Algérie occidentale). *Libyca*, 175–205.

Chamberlain, A.T. and Wood, B.A. (1987). Early Hominid phylogeny. *J. Human Evol.*, 16, 50–68.

Chamla, M.-C. (1968). L'évolution du type de Mechta-Afalou en Algérie occidentale. *C.R. Acad. Sc.*, 267(D), 1949–1951.

Chamla, M.-C. (1970). *Les hommes épipaléolithiques de Columnata (Algérie Occiden-tale) : étude anthropologique*. CRAPE, Algiers.

Chateau, M. (1975). *Orthopédie dento-faciale. Clinique (diagnostic et traitement)*, 5th edition. Julien Prelat, Paris.

Cho, K.-H. (2005). Asymétries cranio-faciales et dysmorphoses. Analyse architecturale cranio-faciale des populations médiévales de La Queue-En-Brie (Val-de-Marne, France). PhD Thesis, Muséum national d'Histoire naturelle, Paris.

Cho, K.-H. and Hadjouis, D. (2005). Les asymétries cranio-faciales de la nécropole médiévale de la Queue-En-Brie (Île-de-France, Val-de-Marne). *Biom. Hum. et Anthrop.*, 23(1/2), 95–104.

Clauzade, M.A. and Darraillans, B. (1989). *Concept ostéopathique de l'occlusion*. SEOO, Perpignan.

Clauzade, M.A. and Darraillans, B. (1992). *L'homme, le crâne, les dents*. SEOO, Perpignan.

Clément-Mullet, J.-J. (2000). *Ibn Al-Awwâm. Le livre de l'agriculture (Kitab al filaha)*. Actes Sud/Sindbad, Arles.

Conroy, G.C., Pickford, M., Senut, B., Van Couvering, J., Mein, P. (1992). *Otavipithecus namibiensis, nov. gen., nov. sp.*, first Miocene Hominoid from Southern Africa (Berg Aukas, Namibia). *Nature*, 356, 144–148.

Coupin, L., Fleury, L., Hadjouis, D. (1995). Premiers résultats des fouilles du Parvis Saint Christophe à Créteil. *Clio*, 94, 4–16.

Couture, C. and Hublin, J.-J. (2005). Les Néandertaliens. In *Origine et évolution des populations humaines*, Dutour, O., Hublin, J.-J., Vandermeersch, B. (eds). Comité des travaux historiques et scientifiques, Paris.

Cregut, E. (1980). *Equus mosbachensis tautavelensis nov. sub. sp.*, un nouvel *Equidae* (*Mammalia*, *Perissodactyla*) du gisement pléistocène moyen de la Caune de l'Arago (Tautavel, Pyrénées orientales, France). *Geobios*, 13(1), 121–127.

Cregut, E. and Bonnoure, E. (1996). Famille des *Hyaenidae*. In *Les Grands Mammifères plio-pléistocènes d'Europe*, Guérin, C., Patou-Mathis, M. (eds). Masson, Paris.

Crubézy, E. and Sellier, P. (1990). Caractères discertes et organisation des ensembles sépulcraux. *Bull. et mém. de la Soc. d'Anthrop. de Paris*, 2(3–4), 171–178.

Dambricourt-Malassé, A. (1987). Ontogenèse, paléontogenèse et phylogenèse du corps mandibulaire catarhinien. Nouvelle interprétation de la mécanique humanisante (théorie de foetalisation, Bolk, 1926). PhD Thesis, Muséum national d'Histoire naturelle, Paris.

Dambricourt-Malassé, A. (1993). Continuity and discontinuity during hominization. *Quaternary Intern*, 19, 85–98.

Dambricourt-Malassé, A. (2004). Un nouvel angle de base pour une meilleure compréhension de la morphogenèse basi-crânienne des hominidés et des modalités évolutives. In *Biométrie & Anthropologie de la tête, de la face et du cou, XIème colloque de la Société de biométrie humaine*. Paris.

Dambricourt-Malassé, A. (2006). La question des origines de la lignée humaine : les modèles évolutionnistes confrontés aux fossiles, implications en orthopédie et posturologie. *Cahiers du CEOPS*, 4, 38–57.

Dambricourt-Malassé, A. (2011). Équilibres bipèdes permanents, origine embryonnaire, morphogenèse, équilibre occluso-postural, conséquences pour l'évolution psychomotrice et comportementale des Hominidés. HDR Thesis, Université de technologie de Compiègne, Compiègne.

Dart, R.A. (1925). *Australopithecus africanus*: The Man-Ape of South Africa. *Nature*, 115, 195–199.

Dastugue, J. (1958). Luxations invétérées du coude sur les squelettes néolithiques. *Bull. Soc. Anthrop. Paris*, 9, 201–209.

Dastugue, J. (1970). Pathologie des hommes de Columnata. In *Les hommes épipaléolithiques de Columnata (Algérie occidentale)*, Chamla, M.-C. (ed.). CRAP, Paris.

Dastugue, J. (1975). Le diagnostic des luxations traumatiques sur les squelettes anciens. *Dos Arquivos de Anatomia E Antropologia*, 1, 221–248.

Dastugue, J. (1994). La luxation traumatique en paléopathologie. Quatre hanches luxées dans un même cimetière médiéval. *Bull. Soc. Anthrop. du Sud-Ouest*, XXIX, 159–177.

Dastugue, J. and Gervais, V. (1992). *Paléopathologie du squelette humain*. Boubée, Paris.

Delaire, J. (1978). L'analyse architecturale et structurale cranio-faciale (de profil). Principes théoriques – quelques exemples d'emploi en chirurgie maxillo-faciale. *Revue de Stomatologie*, 79, 1–33.

Delaire, J. (1980). Essai d'interprétation des principaux mécanismes liant la statique à la morphogenèse céphalique. *Act. Odont. Stom.*, 130, 189–220.

Delattre, P. and Fenart, R. (1960). *L'hominisation du crâne*. CNRS, Paris.

Deshayes, M.-J. (1986). *Croissance cranio-faciale et orthodontie : rapports de biodynamique crânienne*. Masson, Paris.

Deshayes, M.-J. (1991). Reconsidération de la croissance cranio-faciale au cours de la croissance et de l'évolution. Conséquences en Orthopédie dento-maxillo-faciale. *Rev. Orthop. Dento-Facial*, 25, 353–365.

Deshayes, M.-J. (1997). La morphogenèse cranio-faciale. CD-ROM, Sid.

Deshayes, M.-J. (2000). *Repérages crâniens. Conception biodynamique de la morphogenèse cranio-faciale. Ses implications en paléontologie et dans les traitements précoces des dysharmonies dento-faciales*. Cranexplo, Caen.

Donnelly, S.M. (1996). How different are KNM-ER 1470 and KNM-ER 1813? A multivariate comparison using randomization. *Am. J. Phys. Anthrop.*, supplement 22, 22.

Dorst, J. and Dandelot, P. (1976). *Guide des grands mammifères d'Afrique*. Delachaux et Nestlé, Paris.

Duchet Suchaux, G. and Pastoureau, M. (2002). *Le bestiaire médiéval. Dictionnaire historique et bibliographique*. Le Léopard d'Or, Paris.

Dufour, J.-Y. (ed.) (2012). Saint-Maurice, Temple et cimetière huguenot de Charenton. Report, INRAP Centre – Île-de-France.

Dufour, J.-Y. (ed.) (2019). *Les fouilles archéologiques du temple et du cimetière huguenots de Charenton*. Honoré Champion, Paris.

Durbet, G., Rodriguez, P., Badalian, L., Hadjouis, D., Gauthier, A., Laurent, M., Ricard, J.-L., Wattez, J. (1997). Découverte d'un site paléolithique moyen dans les alluvions saaliennes du confluent Seine-Marne à Maisons-Alfort (Val-de-Marne). *Comptes rendus de l'Académie des Sciences*, 324(IIa), 505–512.

Duvernoy, G.L. (1851). Note sur une espèce fossile de Buffle (*Bubalus* (Arni) *antiquus*) découverte en Algérie. *Compte rendu hebdomadaire des séances de l'Académie des Sciences de Paris*, 33, 595–597.

Eisenmann, V. (1977). Les hipparions africains : valeur et signification de quelques caractères des jugales inférieures. *Bull. Mus. nat. Hist. nat.*, 3rd series(438), 69–87.

Eisenmann, V. (1979). Les chevaux (*Equus sensu lato*) fossiles et actuels : étude craniologique et odontologique. PhD Thesis, Université Pierre et Marie Curie, Paris.

Eisenmann, V. (1980). *Les chevaux (*Equus sensu lato*) fossiles et actuels : étude craniologique et odontologique (Cahiers de Paléontologie)*. CNRS, Paris.

Eisenmann, V. (1981). Étude des dents jugales inférieures des *Equus* (*Mammalia, Perissodactyla*) actuels et fossiles. *Paleovertebrata*, 127–226.

Eisenmann, V. (1988). Tentative typology and biostratigraphy of some Middle and Late Pleistocene Western European horses. *Il Quaternario*, 1(2), 13–109.

Eisenmann, V. (1991). Les chevaux quaternaires européens (*Mammalia, Perisso-dactyla*) : taille, typologie, biostratigraphie et taxonomie. *Geobios*, 24(6), 747–759.

Eisenmann, V. (1995). L'origine des ânes : questions et réponses paléontologiques. *Ethnozootechnie*, 56, 5–26.

Eisenmann, V. and Sondaar, P.Y. (1989). Hipparions and the Mio-Pliocene boundary. *Bull. Soc. Paleont. ital.*, 28(2–3), 217–226.

Eisenmann, V., Cregut-Bonnoure, E., Moigne, A.-M. (1985). *Equus mosbachensis* et les grands chevaux de la Caune de l'Arago et de Lunel-Viel : craniologie comparée. *Bull. Mus. Nat. Hist. Nat.*, 4th series(7), 157–173.

Enlow, D.H. (1968). *The Human Face. Hoeber Medical Division*. Harper and Row, New York.

Enlow, D.H. (1973). Growth and the problem of the local control mechanism. *Am. J. Anat.*, 136, 403–406.

Enlow, D.H. (1982). *Handbook of Facial Growth*. Saunders & Co, Philadelphia.

Esclassan, R., Hadjouis, D., Donat, R., Passarius, O., Maret, D., Vaysse, F., Crubézy, E. (2015). A panorama of tooth wear during the medieval period. *J. Biol. Clinic. Anthropol.*, 185–199.

Fenart, R. (1970). L'hominisation du crâne. *Bull. de l'Acad. Dent.*, 14, 33–43.

Ferembach, D. (1962). *La nécropole épipaléolithique de Taforalt (Maroc oriental). Étude des squelettes humains*. CNRS, Casablanca.

Fernandez, P., Bouzouggar, A., Collina-Girard, J., Coulon, M. (2015). The last occurence of *Megaceroides algericus* Lydekker, 1890 (*Mammalia, Cervidae*) during the middle Holocene in the cave of Bizmoune (Morocco, Essaouira region). *Quaternary International*, 374, 154–167.

Fontugne, M., Oujaa, A., Ouchaou, B., Gourari, L., Zouhri, S., Moreau, C., Kaltnecker, E., Dumoulin, J.-P., Werner, M., Benabdelhadi, M. (2012). Sur la présence d'*Ursus arctos* au début de l'Holocène dans le Moyen Atlas (Maroc). Datations et implications paléoenvironnementales. *Quaternaire*, 23(2), 6209.

Gabunia, L., Vekua, A., Lordkipanidze, D., Swisher III, C.C., Ferring, R., Justus, A., Nioradze, M., Tvalchrelidze, M., Antón, S.C., Bosinski, G., Jöris, O., de Lumley, M.-A., Majsuradze, G., Mouskhelishvili, A. (2000). Earliest Pleistocene hominid cranial remains from Dmanisi, Republic of Georgia: Taxonomy, geological setting, and age. *Science*, 288, 1019–1025.

Gabunia, L., Anton, S.C., Lordkipanidze, D., Vekua, A., Justus, A., Swisher, C.C. (2001). Dmanisi and dispersal. *Evolutionary Anthropology*, 10, 158–170.

Gabunia, L., de Lumley, M.-A., Vekua, A., Lordkipanidze, D., de Lumley, H. (2002). Découverte d'un nouvel hominidé à Dmanissi (Transcaucasie, Géorgie). *Comptes rendus palevol.*, 1(4), 243–253.

Garbarino, J. (1979). *La denture de l'homme dans l'appareil masticateur. Anatomie dentaire*. Médiprint, Marseille.

Gaspard, M. (1972). Deux propriétés antagonistes : l'acuité et la robustesse des cuspides. *L'information dentaire*, 125–133.

Gentry, A.W. (1967). *Pelorovis oldowayensis* Reck, an extinct bovid from East Africa. *Bull. British Mus.*, 14, 243–299.

Gentry, A.W. and Gentry, A. (1978a). Fossil *Bovidae* (*Mammalia*) of Olduvai Gorge, Tanzanie. Part I. *Bull. British Mus.*, 29, 289–446.

Gentry, A.W. and Gentry, A. (1978b). Fossil *Bovidae* (*Mammalia*) of Olduvai Gorge, Tanzanie. Part II. *Bull. British Mus.*, 30, 1–83.

Geraads, D. (1981). *Bovidae* et *Giraffidae* (*Artiodactyla, Mammalia*) du Pléistocène de Ternifine (Algérie). *Bulletin du Muséum national d'histoire naturelle de Paris*, 4ème série, 3(1), 47–86.

Geraads, D. (1997). Carnivores du Pliocène terminal d'Ahl Al Oughlam (Casablanca, Maroc). *Geobios*, 30(1), 127–164.

Geraads, D. and Amani, F. (1997). La faune du gisement à *Homo erectus* de l'Aïn Maarouf, près de El Hajeb (Maroc). *L'Anthropologie*, 101(3), 522–530.

Ghetie, B. and Mateesco, C.N. (1977). L'élevage et l'utilisation des Bovins au Néolithique moyen et tardif du Bas-Danube et du Sud des Balkans. *L'Anthropologie*, 81, 115–128.

Ginsburg, L., Hilly, J., Taquet, P. (1968). Une faune würmienne dans un remplissage de fente du Massif de Filfila (littoral nord-constantinois). *Compte rendu sommaire de la Société géologique de France*, 5, 157–158.

Gommery, D. and Senut, B. (2002). *Orrorin tugenensis* distal thumb. In *International Workshop, From Samburupithecus to Orrorin: Origin of Hominids. Geological and Palaeontological Background*. Bogorio, Nigeria.

Grmek, M. (1983). *Les maladies à l'aube de la civilisation occidentale*. Payot, Paris.

Groves, C.P. (1966). Taxonomy. In *Sull'asino selvatico africano*, Groves, C.P., Ziccardi, F., Toschi, A. (eds). *Supplemento alle Ricerche di Zoologia Applicata alla Caccia*, 5(1), 2–11.

Groves, C.P. (1986). The taxonomy, distribution and adaptation of recent equids. In *Equids in the Ancient World*, Meadows, R.H., Uerpmann, H.P. (eds). Ludwig Reichert, Wiesbaden.

Groves, C.P. and Mazak, V. (1975). An approach of taxonomy of the *Hominidae*: Gracile Villafranchian Hominids in Africa. *Casopis pro Mineralogii geologii*, 20, 225–247.

Guadelli, J.-L. (1987). Contribution à l'étude des zoonoses préhistoriques en Aquitaine (Würm ancien et interstade würmien). Thesis, Université de Bordeaux, Bordeaux.

Gudin, R.G. (1951). Étude du crâne de profil : Anthropomorphes et Hominiens. Thesis, Paris.

Gudin, R.G. (1952). Contribution à l'étude du prognathisme ethnique par la téléradiographie. Thesis, Académie nationale de médecine, Paris.

Gudin, R.G. (1954). Étude du crâne de profil. Morphologie et radiographie. *Bulletins et Mémoires de la Société d'anthropologie de Paris*, 3.

Gudin, R.G. (2002). *L'homme dans son évolution. Bipédie et langage. L'architectonique cranio-faciale par la téléradiographie*. Les Études Hospitalières, Bordeaux.

Guemriche, S. (2007). *Dictionnaire des mots français d'origine arabe*. Le Seuil, Paris.

Guérin, C. (1994). *L'homme et la domestication des animaux*. ARPAM/Muséum de Lyon, Lyon.

Guérin, C. (1996a). Super-ordre des Proboscidiens. In *Les grands mammifères plio-pléistocènes d'Europe*, Guérin, C., Patou-Mathis, M. (eds). Masson, Paris.

Guérin, C. (1996b). Ordre des Périssodactyles. In *Les grands mammifères plio-pléistocènes d'Europe*, Guérin, C., Patou-Mathis, M. (eds). Masson, Paris.

Guérin, C. and Patou-Mathis, M. (eds) (1996). *Les grands mammifères plio-pléistocènes d'Europe*. Masson, Paris.

Hadjouis, D. (1982). Les Canidés de la grotte du Lazaret (Alpes-Maritimes). Étude paléontologique. DEA Thesis, Université Paris 6, Muséum national d'Histoire naturelle, Paris.

Hadjouis, D. (1985a). Les Bovidés du gisement atérien des Phacochères (Alger, Algérie). Contribution à l'étude des Bovidés du Pléistocène moyen et supérieur du Maghreb. Thesis, Université Paris 6, Muséum national d'Histoire naturelle, Paris.

Hadjouis, D. (1985b). Les Bovidés du gisement atérien des Phacochères (Alger). Interprétations paléoécologiques et phylogénétiques. *Comptes rendus de l'Académie des Sciences*, 301(II), 1251–1254.

Hadjouis, D. (1990). *Megaceroides algericus* (Lydekker, 1890), du gisement des Phacochères (Alger, Algérie). Étude critique de la position systématique de *Megaceroïdes*. *Quaternaire*, 3(4), 247–258.

Hadjouis, D. (1994). Taphonomie des faunes du gisement atérien des Phacochères (Alger, Algérie). In *Outillage peu élaboré en os et en bois de Cervidés, IV : taphonomie/bone modification*, Patou-Mathis, M. (ed.). Éditions du Centre d'études et de documentation archéologiques, Treignes.

Hadjouis, D. (1998a). Le cheval de Maisons-Alfort (Val-de-Marne, France) et ses rapports avec les chevaux du Pléistocène moyen et supérieur d'Europe. *Quaternaire*, 9(4), 337–343.

Hadjouis, D. (1998b). Les relations entre les pathologies du rachis cervical et les dysharmonies cranio-faciales et dentaires. Applications téléradiographiques aux populations médiévales du Val-de-Marne et reconstitutions faciales. *IVème colloque de la Soc. de Biom. Hum.*, 16(1–2), 49–58.

Hadjouis, D. (1999). *Les populations médiévales du Val-de-Marne. Dysharmonies cranio-faciales, maladies bucco dentaires et anomalies du développement dentaire au cours du Moyen Âge*. Artcom', Paris.

Hadjouis, D. (2000a). Équilibre ou déséquilibre, normoclusion ou malocclusion. Actualités Odonto-Stomatologiques. *L'Encyclopédie du Praticien*, 213, 35–45.

Hadjouis, D. (2000b). Étiopathogénies des dysharmonies cranio-faciales de populations médiévales du Val-de-Marne. In *Actes du Colloque de Créteil*, Andrieux, P., Hadjouis, D., Dambricourt-Malassé, A. (eds). Artcom', Paris.

Hadjouis, D. (2001). Proposition d'une fiche d'analyse en craniologie et en paléo-odontologie. In *La Paléo-odontologie. Analyses et méthodes d'étude*, Hadjouis, D., Mafart, B. (eds). Artcom', Paris.

Hadjouis, D. (2002). Un nouveau *Bovini* dans la faune du Pléistocène supérieur d'Algérie. *L'Anthropologie*, 106, 377–385.

Hadjouis, D. (2003a). La faune mammalienne d'Algérie, un renouvellement constant. *Préhistoire de l'Algérie. Dossiers d'Archéologie*, 282(April).

Hadjouis, D. (2003b). Les hommes modernes d'Algérie, origine et paléogéographie. *Préhistoire de l'Algérie. Dossiers d'Archéologie*, 282(April).

Hadjouis, D. (2003c). L'avulsion dentaire : une pratique, plusieurs effets. *Préhistoire de l'Algérie. Dossiers d'Archéologie*, 282 (April).

Hadjouis, D. (2003d). Hominidés et Grands mammifères dans leur contexte paléoenvironnemental au cours du Quaternaire maghrébin. HDR Thesis, Université de Perpignan, Perpignan.

Hadjouis, D. (2005). *Pelorovis antiquus* et/ou *Syncerus caffer* ? *Taurotragus* ou *Tragelaphus* ? Les buffles et les *Tragelaphini* du gisement atérien des Phacochères (Alger, Algérie). *Quaternaire*, Special edition (5), 91–94.

Hadjouis, D. (2006a). L'homme face aux maladies du passé. Catalogue d'exposition, La Science et nous, Val-de-Marne.

Hadjouis, D. (2006b). Variations anatomiques posturales chez les populations franciliennes du Moyen Âge (Val-de-Marne). L'exemple des dysplasies luxantes de la hanche et des malformations articulaires du genou. *Biom. Hum. et Anthrop.*, 24, 3–4.

Hadjouis, D. (2008a). L'industrie osseuse de Maisons-Alfort (ZAC d'Alfort 2). Corpus des outillages en matière dure animale du Néolithique ancien au Bronze final. Zooarachaeology Report, laboratoire départemental d'Archéologie, Conseil général du Val-de-Marne.

Hadjouis, D. (2008b). Déséquilibre architectural cranio-facial et ses relations avec l'occlusion et le squelette post-crânien. In *Ostéo-archéologie et techniques médico-légales : tendances et perspectives. Pour un "Manuel pratique de paléo-pathologie humaine"*, Charlier, P. (ed.). De Boccart, Paris.

Hadjouis, D. (2009). Les maladies alvéolo-dentaires chez les populations médiévales du Val-de-Marne. *Société française d'histoire de l'art dentaire*, 14, 38–42.

Hadjouis, D. (2011). *Les hommes de Mechta-Afalou d'Algérie. Architecture cranio-faciale, occlusion biodynamique, paléopathologie. L'exemple des hommes paléolithiques d'Afalou-Bou-Rhummel.* CNRPAH, Algiers.

Hadjouis, D. (2012a). Étude paléopathologique et épidémiologique. Excavation report, Saint-Maurice, temple et cimetière huguenot de Charenton, Dufour, J.-Y (ed.).

Hadjouis, D. (2012b). La craniosténose des populations médiévales du Val-de-Marne. Poster, International Symposium. Ostéopathie et Transdisciplinarité, le squelette humain dans tous ses états. École supérieure d'ostéopathie de Paris – Marne-la-Vallée, 24–26 May.

Hadjouis, D. (2012c). Présence de la tuberculose articulaire sur un squelette complet du XIVème siècle à Villejuif (nécropole de l'église Saint-Cyr/Sainte-Julitte, Val-de-Marne). Poster at, International Symposium Ostéopathie et Transdisciplinarité, le squelette humain dans tous ses états. École supérieure d'ostéopathie de Paris – Marne-la-Vallée, 24–26 May.

Hadjouis, D. (2012d). Scoliose congénitale par défaut segmentaire sur un squelette de femme du XVIIème siècle (Saint-Maurice-Val de-Marne). Poster, International Symposium Ostéopathie et Transdisciplinarité, le squelette humain dans tous ses états. École supérieure d'ostéopathie de Paris – Marne-la-Vallée, 24–26 May.

Hadjouis, D. (2012e). Occlusion et biodynamique cranio-faciale des populations modernes du Bassin parisien. Poster, International Symposium Ostéopathie et Transdisciplinarité, le squelette humain dans tous ses états. École supérieure d'ostéopathie de Paris – Marne-la-Vallée, 24–26 May.

Hadjouis, D. (2014). Équilibre postural des Hominidés d'Algérie et du Maghreb et leurs relations architecturales cranio-faciales avec les populations contemporaines. *Ikosim*, 3, 8–20.

Hadjouis, D. (2015a). Peuplements de l'Afrique du Nord. *L'encyclopédie berbère*, 37, 6034–6052.

Hadjouis, D. (2015b). Paléopathologie des Hominidés de l'Afrique du Nord. *L'encyclopédie berbère*, 37, 6053–6066.

Hadjouis, D. (2016a). Vers un paradigme holistique de l'anatomie paléo-anthropologique : entre adaptation posturale et globalité architecturale du corps chez les Hominidés. *Ikosim*, 7–20.

Hadjouis, D. (2016b). Qui sont les mammouths d'Île-de-France ? *Archéologia*, 548, 46–51.

Hadjouis, D. (2017a). Un geste chirurgical bucco-dentaire datant du Moyen Âge. *L'information dentaire*, 13(99), 42–43.

Hadjouis, D. (2017b). Forte fréquence de malformations architecturales cranio-faciales et de classe III dans une population médiévale du sud-est parisien (La Queue-en-Brie, Val-de-Marne). *L'Orthodontie bioprogressive*, 8–11.

Hadjouis, D. (2018). *Atlas des maladies et traumatismes du monde médiéval et moderne : Ve siècle-XVIIe siècle*. ISTE Editions, London.

Hadjouis, D. (2019a). *Syncerus* et les buffles en Afrique du Nord. *L'encyclopédie berbère*, 38.

Hadjouis, D. (2019b). Paléopathologie des squelettes de Charenton. In *Les fouilles archéologiques du temple et du cimetière huguenots de Charenton*, Dufour, J.Y. (ed.). Honoré Champion, Paris.

Hadjouis, D. (2020a). Les mammouths du Val-de-Marne. Catalogue d'exposition, Musée Archéa, Louvres.

Hadjouis, D. (2020b). Paléopathologie des squelettes de Chennevières-sur-Marne. In *Rapport final de synthèse*, Bayard, V. (ed.). Service archéologie du Val-de-Marne, Villejuif.

Hadjouis, D. (2020c). La posture occlusale des populations juvéniles du Maghreb et du Sahara depuis le Paléolithique supérieur jusqu'à l'actuel. In *Journées scientifiques de l'Université de la Laguna, Circulations durant l'Holocène autour du Maghreb*. Tenerife, 17–21 February.

Hadjouis, D. and Katz, P. (1998). Craniosténoses du vivant et des populations du Moyen Âge. Analyse téléradiographique comparée des troubles cranio-faciaux et dentaires et reconstitution faciale. In *IVème colloque de Biométrie humaine, Biométrie maxillo-faciale et dentaire*, 16(3–4), 179–187.

Hadjouis, D. and Katz, P. (2020). The importance of paralleling technique in the diagnostic of dental-skeletal dysmorphia in ancient populations. Cranio-facial architecture and occlusion. *Journ. of Forens. Radiol. and Imag.*, 20(2020), 100344.

Hadjouis, D. and Le Bihan, P. (2014). Atlas des mammifères quaternaires et actuels d'Algérie. *Travaux du CNRPAH*, 20.

Hadjouis, D. and Mafart, B. (eds) (2001). *La Paléo-Odontologie. Analyses et méthodes d'étude*. Artcom', Paris.

Hadjouis, D. and Roubet, C. (2020). Le site atérien des Phacochères à Hydra (Alger, Algérie) : les actions anthropiques et prédatrices dans le Sahel d'Alger. *Bull. Archéol. Maroc. Rabat*, forthcoming.

Hadjouis, D. and Sahnouni, M. (2006). *Pelorovis howelli nov. sp.* (*Mammalia, Artiodactyla*): A new bovine from the lower Pleistocene of Aïn Hanech (El-Kherba locus), Northern Algeria. *Geobios*, 39, 673–678.

Hadjouis, D. and Thillaud, P. (1994). Alternatives to an identification of calcified pleural plaques: Concerning two cases dated VI-VIII century (France). *Homo*, 45, S55.

Hadjouis, D., Vignal, J.-N., Schuliar, Y. (1997). Imagerie numérisée et reconstitution faciale. *Archéologia*, 340, 48–57.

Hadjouis, D., Vignal, J.-N., Medig, M., Sahnouni, M., Derradji, A. (2000a). Bio-dynamique cranio-faciale, paléopathologie et reconstitution faciale 2D d'un crâne ibéromaurusien (grotte de Taza 1, Jijel, Algérie). In *Actes du colloque de Créteil*, Andrieux, P., Hadjouis, D., Dambricourt-Malassé, A. (eds). Artcom', Paris.

Hadjouis, D., Vignal, J.-N., Michaut, J.-F., Richebé, J., Schuliar, Y. (2000b). Analyse architecturale, paléopathologie et reconstitution faciale 2D et 3D (sculpture) d'un crâne mechtoïde du Néolithique de l'Ouest algérien (Oued Guettara, Algérie). In *Actes du colloque de Créteil*, Andrieux, P., Hadjouis, D., Dambricourt-Malassé, A. (eds). Artcom', Paris.

Hadjouis, D., Corbineau, R., Ruas, M.P., Verdin, P. (2009). Techniques d'embaumement sur le corps d'un noble anglais mort de la peste (Saint Maurice, Val de Marne). In *Actes du 3ème colloque international de pathographie*. Théâtre de Bourges, Bourges, 3–5 April 2008.

Hadjouis, D., Djennabi, A., Pittion, C., Mittard, P., Bailly, I., Huard, P. (2015a). La nécropole de l'église Saint-Cyr/Sainte Julitte de Villejuif (Val-de-Marne) : population et état sanitaire au cours du Bas Moyen Âge. *Revue archéologique d'Île-de-France*, supplement 3, 233–239.

Hadjouis, D., Huard, P., Lichon, A.-A., Pittion, C. (2015b). La nécropole de l'église Sainte-Colombe de Chevilly-Larue (Val-de-Marne) : population et état sanitaire du Haut et Bas Moyen Âge. *Revue archéologique d'Île-de-France*, supplement 3, 241–249.

Hadjouis, D., Bailly, I., Huard, P. (2020). Corps de militaires retrouvés au fort de Kremlin-Bicêtre et autres squelettes de la guerre de 1870 dans le Val-de-Marne. In *Actes du colloque archéologique d'Île-de-France*. Créteil, November 2018.

Hailé-Sélassié, Y. (2000). Late Miocene Hominids from the middle Awash, Ethiopia. *Nature*, 412, 178–181.

Hamdine, W., Thévenot, M., Michaux, J. (1998). Histoire récente de l'ours brun au Maghreb. *Acad. Sc. Elsevier*, 566–570.

Harris, J.M. and White, T. (1979). Evolution of the Plio-Pleistocene African Suidae. *Transactions of the American Philosophical Society*, 69, 1–128.

Harrison, T. (2002). Late oligocene to middle Miocene catarrhines from Afro-Arabia. In *The Primate Fossil Record*, Hartwig, W.C. (ed.). Cambridge University Press, Cambridge.

Heim, J.-L. (1976). *Les hommes fossiles de la Ferrassie, tome I : le gisement, les squelettes adultes (crâne et squelette du tronc)*. Les Archives de l'Institut de Paléontologie Humaine, Paris.

Heim, J.-L., Hadjouis, D., Aumassip, G. (2011). L'enfant néolithique (*Homo* 5) de Tin Hanakaten (Tassili des Ajjers, Algérie). Examen anthropologique et paléopathologique. In *Actes du Colloque international de Préhistoire maghrébine*, 11(2), 27–40.

Heintz, E. (1970). Les Cervidés villafranchiens de France et d'Espagne. *Mémoires du Muséum national d'Histoire naturelle*, 22(C), 1–303.

Hendey, Q.B. (1968). The Melkbos site: An upper Pleistocene fossil occurrence in the South-Western Cape province. *Ann. South Africa Mus. Cape Town*, 52, 89–119.

Hopwood, A.T. (1933). Miocene primates from Kenya. *J. Linnean Soc.*, 38, 437–464.

Huass, H. and Petter, F. (1900). *Les Mammifères du monde entier*. Nathan, Paris.

Hublin, J.-J. (1978). Quelques caractères apomorphes du crâne néandertalien et leur interprétation phylogénique. *Comptes rendus de l'Académie des Sciences*, 291, 669–679.

Hublin, J.-J. (1988). Caractères dérivés de la région occipito mastoïdienne chez les Néandertaliens. In *L'Homme de Néandertal, tome 3 : l'anatomie*, Trinkaus, E. (ed.). Éraul, Liège.

Hublin, J.-J. (2005). D'*Homo ergaster* à *Homo sapiens*. In *Origine et évolution des populations humaines*, Dutour, O., Hublin, J.-J., Vandermeersch, B. (eds). Comité des travaux historiques et scientifiques, Paris.

Hublin, J.-J., Ben-Ncer, A., Bailey, S.E., Freidline, S.E., Neubaur, S., Skinner, M.M., Bergmann, I., Le Cabec, A., Benazzi, S., Harvati, K., Gunz, P. (2017). New fossils from Jebel Irhoud, Morocco ant the pan-African origin of *Homo sapiens*. *Nature*, 546, 289–292.

Ishida, H. and Pickford, M. (1998). A new late Miocene Hominoid from Kenya: *Samburupithecus kiptalami, gen.* et *sp. nov. Comptes rendus de l'Académie des Sciences*, IIa, 325, 823–829.

Ishida, H., Pickford, M., Nakaha, H., Nakano, Y. (1984). Fossil anthropoids from Nachola and Samburu Hills. *Afr. Stud. Mon.*, 2, 73–85.

Jamet, E. (2000). Trépanation crânienne au Néolithique dans le Bassin parisien. In *L'identité humaine en question, Actes du colloque de paléontologie et de paléoanthropologie de Créteil*, Andrieux, P., Hadjouis, D., Dambricourt Malassé, A. (eds). Artcom', Paris.

Johanson, D.C., White, T., Coppens, Y. (1978). A new species of the genus *Australopithecus* (Primates: *Hominidae*) from the pliocene of Eastern Africa. *Kirtlandia*, 28.

Joleaud, L. (1913). Études de géographie zoologique de la Berbérie, I : les Cervidés. *Revue africaine*, 56, 471–499.

Joleaud, L. (1914). *Cervus* (*Megaceroïdes*) *algericus* Lydekker 1890. *Compte rendu de la Société biologique*, 76, 737–738.

Joleaud, L. (1916). *Cervus* (*Megaceroïdes*) *algericus* Lydekker 1890. *Rec. et Not. de la Soc. Archéol. Constantine*, 49, 1–67.

Kahle, W., Leonhardt, H., Platzer, W. (1980). *Anatomie, tome 1 : appareil locomoteur*. Flammarion, Paris.

Korpal, Y. (2005). L'os sphénoïde dans la lignée humaine. PhD Thesis, Muséum national d'Histoire naturelle, Paris.

Kroitor, R. (2016). Systematic position and paleoecology of the endemic deer *Megaceroides algericus* Lydekker, 1890 (*Cervidae*, *Mammalia*) from Late Pleistocene-early Holocene of North Africa. *Geobios*, 49, 265–283.

Kroitor, R. and Bonifay, M.-F. (2001). Étude préliminaire des cerfs du gisement Pléistocène inférieur de Ceyssaguet (Haute-Loire). *Paleo*, 13, 129–153.

Kroitor, R., Bonifay, M.-F., Bonifay, E. (2006). Origin and evolution of the late Pleistocene island deer *Praemegaceros* (*Nesoleipoceros*) cazioti (Depéret) from Corsica and Sardinia. *Bulletin du Musée d'anthropologie préhistorique de Monaco*, 46, 35–68.

Kurtén, B. (1963). The cave hyena, an essay in statistical analysis. In *Science in Archeology*, Brothwell, R., Higgs, E.S. (eds). Thames & Hudson, London.

Kurtén, B. (1968). *Pleistocene Mammals of Europe*. Weidenfeld and Nicholson, London.

Kurtén, B. and Anderson, E. (1980). *Pleistocene Mammals of North America*. Columbia University Press, New York.

Kurtén, B. and Poulianos, N. (1981). Fossil Carnivora of Petralona cave, status of 1980. *Anthropos*, 8, 9–56.

Langlade, M. (1981). *Diagnostic orthodontique*. Maloine S.A., Paris.

Le Double, A.F. (1903). *Traité des variations des os du crâne de l'homme et leur signification au point de vue de l'Anthropologie zoologique*. Vigot-Maloine, Paris.

Le Double, A.F. (1906). *Traité des variations des os de la face de l'homme et leur signification au point de vue de l'Anthropologie zoologique*. Vigot-Maloine, Paris.

Le Gros Clark, W.E. (1964). *The Fossil Evidence for the Human Evolution*. Chicago University Press, Chicago.

Le Gros Clark, W.E. and Leakey, L.S.B. (1950). Diagnoses of East African Miocene Hominoidea. *Quat. J. Geol. Soc. Lond.*, 105, 260–262.

Leakey, L.S.B. (1967). Notes on the mammalian faunas from the Miocene and Pleistocene of East Africa. In *Background to Evolution in Africa*, Bishop, W.W., Clark, J.D. (eds). University of Chicago Press, Chicago.

Leakey, R.E.F. (1976). New Hominids fossils from Koobi Fora formation in northern Kenya. *Nature*, 261, 574–576.

Leakey, L.S.B., Tobias, P.V., Napier, J.R. (1964). A new species of the genus *Homo* from Olduvay Gorge. *Nature*, 202, 7–9.

Leakey, M.G., Feibel, C.S., McDougall, I., Ward, C., Walker, A. (1998). New specimens and confirmation of an early age for *Australopithecus anamensis*. *Nature*, 393, 62–66.

Lecointre, G. and Le Guyader, H. (2001). *Classification phylogénétique du vivant*, 2nd edition. Belin, Paris.

Lemoine, P. and Teilhard de Chardin, P. (1923). Les gros blocs quaternaires du port de Bonneuil. *La Nature*, 2560, 272.

Leroi-Gourhan, A. (1983). *Le geste et la parole : technique et langage*. Albin Michel, Paris.

Lieberman, D.E., Pilbeam, D.R., Wood., B.A. (1988). A probalistic approach to the problem of sexual dimorphism in *Homo habilis*: A comparison of KNM-ER 1470 and KNM-ER 1813. *J. Hum. Evol.*, 17, 503–511.

Lieberman, D.E., Wood., B.A., Pilbeam, D.R. (1996). Homoplasy and early *Homo*: An analysis of the evolutionary relationships of *Homo habilis* sensu stricto and *Homo rudolfensis*. *J. Hum. Evol.*, 30, 97–120.

Lonnberg, E. (1933). Description of a fossil Buffalo from East Africa. *Arkiv Zool. Stockholm*, 25(17), 1–32.

Lordkipanidze, D. (2008). Étonnants primitifs de Dmanissi. *La Recherche*, 419, 28–32.

Lydekker, R. (1890). On a Cervine jaw from Algeria. *Proceed. Zool. Soc.*, 602–604.

Malbot, H. and Verneau, R. (1897). Les Chaouias et la trépanation du crâne dans l'Aurès. *L'Anthropologie*, 8.

Manouvrier, L. (1903). Les marques sincipitales des crânes néolithiques considérées comme reliant la chirurgie classique à la chirurgie préhistorique. *Bull. Soc. Anthrop. Paris*, 2.

Manzi, G., Mallegni, F., Ascenzi, A. (2001). A cranium for the earliest Europeans: Phylogenetic position of the Hominid from Ceprano. *Proceedings of the National Academy of Science*, 98, 10011–10016.

Maroteaux, P. (1995). *Les maladies osseuses de l'enfant*, 3rd edition. Flammarion, Paris.

Martin, L.T. (1867). *La trépanation du crâne telle qu'elle est pratiquée par les kabyles de l'Aurès*. Montpellier médical, Montpellier.

Michaux, J. and Thevenot, M. (2007). Histoire récente de l'ours brun (Ursus arctos Linné, 1758) au Maghreb. Contribution des fossiles d'Algérie. *Mésogée*, 63, 43–49.

Michel, P. (1990). Contribution à l'étude paléontologique des Vertébrés fossiles du quaternaire marocain à partir de sites du Maroc atlantique, central et oriental. PhD Thesis, Musée national d'Histoire naturelle, Paris.

Moss, M.L. (1969). The primary role of the functional matrix in facial growth. *Am. J. Orthod.*, 55, 566–577.

Moss, M.L. (1974). Vertical growth of the human face. *Am. J. Orthod.*, 50–125, 359–376.

Mourer-Chauviré, C. (1972). Étude de nouveaux restes de vertébrés provenant de la carrière. Fournier à Chatillon-Saint-Jean (Drôme) III. Artiodactyles, chevaux et oiseaux. *Bull. Assoc. fr. Et. Quat.*, 4, 271–305.

Mourer-Chauviré, C. (1980). Le gisement pléistocène supérieur de la grotte de Jaurens à Nespouls (Corrèze, France). Les Équidés (*Mammalia, Perissodactyla*). *Nouv. Arch. Mus. Hist. Nat. Lyon*, 18, 17–60.

Nakatsukasa, M., Yamanaka, A., Kunimatsu, Y., Shimizu, D., Ishida, H. (1998). A newly discovered *Kenyapithecus* skeleton and its implications for the evolution of positional behavior in Miocene East African Hominoids. *J. Hum. Evol.*, 34, 657–664.

Oboussier, H. and Ernst, D. (1977). Der formative Einfluss des Lebensraumes auf das posteranial Skelett der Tragelaphineu (Tribus *Tragelaphini* Sokolov, 1953, *Bovidae, Mammalia*). *Zool. Jahrb. Syst. Iena*, 104, 203–238.

Olive, F. (2005). Évolution des grands Carnivores pliocènes et pléistocène de l'Ancien Monde. Systématique, biochronologie, paléogéographie et paléoécologie. PhD Thesis, Muséum national d'Histoire naturelle, Paris.

Ouchaou, B. and Amani, F. (2002). Les Carnivores des gisements néolithiques et protohistoriques du nord du Maroc. *Quaternaire*, 13(1), 79–87.

Paris, A. (1867). De la trépanation céphalique pratiquée par les médecins de l'Aouress (province de Constantine). *Gazette médicale d'Algérie*, 13(3), 25–28.

Patterson, B. and Howells, W.W. (1967). Hominid humeral fragment from early Pleistocene of Northwestern Kenya. *Science*, 156, 64–66.

Petter, G. and Howell, F.C. (1987). *Machairodus africanus* Arambourg, 1970 (*Carnivora, Mammalia*) du Villafranchien d'Aïn Brimba, Tunisie. *Bulletin du Muséum national d'Histoire naturelle*, 4th series (1), 97–119.

Philippe, J. (2008). Évolution de l'Orthodontie. In *40 ans de chirurgie dentaire, 1968–2008*. Privat, Toulouse.

Pickford, M. and Ishida, H. (1998). Interpretation of *Samburupithecus*, an Upper Miocene Hominoid from Kenya. *Comptes rendus de l'Académie des Sciences*, IIa, 326, 299–306.

Pickford, M. and Senut, B. (2001). The geological and faunal context of Late Miocene Hominid remains from Lukeino, Kenya. *Comptes rendus de l'Académie des Sciences*, IIa, 332, 145–152.

Pickford, M., Senut, B., Conroy, G.C., Mein, P. (1994). Phylogenetic position of *Otavipithecus*: Questions of methodology and approach. In *Current Primatology, I: Ecology and Evolution*, Thierry, B., Anderson, J.R., Roeder, J.J., Herrenschmitd, N. (eds). University Louis Pasteur, Strasbourg.

Piveteau, J. (1973). *Origine et destinée de l'homme*. Masson, Paris.

Pomel, A. (1892). Sur deux ruminants de l'époque néolithique en Algérie : *Cervus pachygenis* et *Antilope maupasi*. *Comptes rendus de l'Académie des Sciences*, 115, 213.

Pomel, A. (1893). *Monographie des Vertébrés fossiles de l'Algérie. Paléontologie. I. Bubalus antiquus. II. Caméliens et Cervidés*. Service de la Carte géologique de l'Algérie, Algiers.

Pomel, A. (1894). *Monographie des Vertébrés fossiles de l'Algérie. III. Bœufs-Taureaux. IV. Les Bosélaphes Ray*. Service de la Carte géologique de l'Algérie, Algiers.

Pouillon, F. (2008). Bêtes médiévales et familiarité : animaux familiers de l'esprit, animaux familiers de la vie. *Anthropozoologica*, 20, 11–40.

Pouillon, F., El-Kader, E., Daumas, E. (2008). *Dialogues sur l'hippologie arabe : les chevaux du Sahara et les mœurs du désert*. Actes Sud, Paris.

Prat, F. (1968). Recherches sur les Équidés pléistocènes en France. PhD Thesis, Faculté des Sciences de Bordeaux, Bordeaux.

Prat, F. (1977). L'équidé du gisement acheuléen de Torralba (Sora, Espagne) *Equus caballus torralbae nov. sub. sp. Bull. AFEQ.*, 50, 33–46.

Prat, F. (1980). *Les Équidés villafranchiens en France : genre Equus*. CNRS, Bordeaux.

Prat, S. (1997). Problème taxinomique des premiers représentants du genre *Homo*. Études crâniennes des individus d'Olduvai et de Koobi Fora. *Bull. Mém. Soc. Anthrop. Paris*, 9(3–4), 251–266.

Prat, S. (2000). Origine et taxinomie des premiers représentants du genre *Homo*. PhD Thesis, Université de Bordeaux I, Bordeaux.

Prat, S. and Marchal, F. (2005). Les premiers représentants du genre *Homo*. In *Origine et évolution des populations humaines*, Dutour, O., Hublin, J.-J., Vandermeersch, B. (eds). Comité des travaux historiques et scientifiques, Paris.

Reck, H. (1928). *Pelorovis oldowayensis, n.g. n.sp.* Wiss. Ergebn. Oldoway-Exped., Leipzig. *N.F.*, 3, 57–67.

Ricketts, R.M. (1960). Cephalometric Synthesis. *Amer. Journ. of Orthodontics*, 46, 647–673.

Ricketts, R.M. (1975). A four step method to distinguish orthodontic changes from natural growth. *J. Clin. Orth.*, 9, 4.

Ricketts, R.M. (1977). The discovery of a law of arcial growth of the mandible. In *Proceedings of the Foundation for Orthodontic Research*. Canoga Park, Los Angeles.

Rightmire, G.P. (1993). Variation among early *Homo* crania from Olduvai Gorge and the Koobi Fora region. *Am. J. Phys. Anthrop.*, 90, 1–33.

Rosas, A. and Bermudez de Castro, J.-M. (1999). The ATD6-5 mandibular specimen from Gran Dolina (Atapuerca, Spain). Morphological study and phylogenetic implica-tions. *J. Hum. Evol.*, 37, 567–590.

Ryckewaert, A. (1987). *Rhumatologie. Pathologie osseuse et articulaire*. Flammarion, Paris.

Sahnouni, M., Parés, J.M., Duval, M., Cáceres, I., Harichane, Z., Van der Made, J., Pérez-González, A., Abdessadok, S., Kandi, N., Derradji, A., Medig, M., Boulaghraif, K., Semaw, S. (2018). 1.9-million and 2.4-million-year-old artifacts and stone tool-cutmarked bones from Ain Boucherit, Algeria. *Science*, 1297–1301.

Sam, Y. (2018). Révision des Équidés (*Mammalia*, *Perissodactyla*) du Pléistocène moyen du lac Karâr (Tlemcen, Algérie). *Geodiversitas*, 40, 171–182.

Sam, Y. (2020). African origins of modern asses as seen from paleontology and DNA: What about the Atlas wild ass? *Geobios*, 73–84.

Sassouni, V.A. (1969). Classification of skeletal facial types. *Amer. Journ. of Orthod.*, 55(2), 109–123.

Sawada, Y., Pickford, M., Itaya, T., Makinouchi, T., Tateishi, M., Kabeto, K., Ishida, S. (1997). K/Ar ages of Miocene Hominoidea (*Kenyapithecus* and *Samburupithecus*) from Samburu Hills, Northern Kenya. *Comptes rendus de l'Académie des Sciences*, IIa(326), 445–451.

Sawada, Y., Murat, T., Pickford, M., Senut, B., Itaya, T., Kashine, C., Hyodo, M., Chujo, T., Fujii, H. (2002). The age of *Orrorin tugenensis*, an early Hominid from the Tugen Hills, Kenya. *C. R. Palevol.*, 1, 293–303.

Saxon, E.C. (1975). The prehistoric economies of the Israeli and Algerian littorals 18000–8000 BP. PhD Thesis, Cambridge University, Cambridge.

Seeley, H.G. (1891). On *Bubalus bainï* (Seeley). *Geological Magazine, London Serv.*, 3(8), 199–202.

Senut, B. (1989). *Le coude des Primates Hominoïdes : anatomie, fonction, taxonomie, évolution*. CNRS, Paris.

Senut, B. (2005). Les Grands Singes miocènes et l'origine des Hominidés. In *Origine et évolution des populations humaines*, Dutour, O., Hublin, J.-J., Vandermeersch, B. (eds). Comité des travaux historiques et scientifiques, Paris.

Senut, B. and Gommery, D. (1997). Squelette post-crânien d'*Otavipithecus*, *Hominoidea* du Miocène moyen de Namibie. *Ann. Paléont. (Vert. Invert.)*, 83(3), 267–284.

Senut, B., Pickford, M., Gommery, D., Kunimatsu, Y. (2000). A new genus of Early Miocene Hominoid from East Africa: *Ugandapithecus major* (Le Gros Clark and Lealey, 1950). *Comptes rendus de l'Académie des Sciences*, IIa, 227–233.

Serre, F. and Hadjouis, D. (1989). Étude des deux premières vertèbres cervicales des *Bovini* (*Bos*, *Bison*, *Pelorovis*). *Revue de Paléobiologie*, 8(1), 163–186.

Shaw, J.C.M. (1939). Growth changes and variations in warthog third molars and their paleontological importance. *Trans. Roy. Soc. South Africa*, 27, 51–94.

Simpson, G.G. (1941). Large Pleistocene felines of North America. *Am. Mus. Novitates*, 1136, 1–27.

Skinazi, G. (1995). *L'orthodontie*. PUF, Paris.

Sonan, N.K., Beugre, J.B., Beugre-Kouassi, A.M.L., Aka, A., Kouadio, M., Diakite, K., Djaha, K. (2006). Variabilité céphalométrique d'une anomalie occlusale à l'intérieur de 3 groupes ethniques différents (mélano-africains, maghrébins et leuco-européens). *Biom. Hum. et Anthropol.*, 24(3–4), 171–181.

Stringer, C.B. (1986). The credibility of *Homo habilis*. In *Major Topics in Primate and Homo Evolution*, Wood, B.A., Martin, L., Andrews, P. (eds). Cambridge University Press, New York.

Sutherland, W.G. (1939). *The Cranial Bowl*. The Free Press, Mankato.

Templeton, A. (2002). Out of Africa again and again. *Nature*, 416(6876), 45–51.

Texier, J.P., Raynal, J.P., Lefèvre, D. (1994). Contribution pour un nouveau cadre stratigraphique des formations littorales quaternaires de la région de Casablanca (Maroc). *Comptes rendus de l'Académie des Sciences*, II(318), 1247–1253.

Thillaud, P.-L. (1996). *Paléopathologie humaine. Traités pratiques d'Archéologie*. Kronos, Sceaux.

Thillaud, P.-L. (2004). Les variations anatomiques d'Anatole-Félix le Double (1848–1913), avant-propos. In *Traité des variations des os du crâne et leur significa-tion au point de vue de l'anthropologie zoologique*, Le Double, A.-F. (ed.). Vigot Maloine, Paris.

Thillaud, P.-L. (2008). Pour une nouvelle approche de l'identification et de l'interprétation des diverses formes de *Cribra orbitalia*. *Histoire des Sciences médicales*, 42(1), 49–62.

Thomas, P. (1875). Ossements du *Bubalus antiquus* découverts à Djelfa en Algérie. *Journal de Zoologie*, 4, 72–78.

Thomas, H. (1977). *Géologie et paléontologie du gisement acheuléen de l'Erg Tihodaïne (Ahagar, Sahara algérien), algérien*. SNED, Algiers.

Thomas, H. and Senut, B. (1999). *Les Primates, ancêtres de l'homme*. Artcom', Paris.

Tillier, A.-M. (1988). À propos des séquences phylogéniques et ontogéniques chez les Néandertaliens. In *L'Homme de Néandertal. 3. L'anatomie*, Trinkaus, E. (ed.). Eraul, Liège.

Tobias, P.V. (1967). *Olduvai Gorge. 2. The Cranium and Maxillary Dentition of Australopithecus (Zinjanthropus) boisei*. Cambridge University Press, Cambridge.

Trinkaus, E. (1983). Neandertal postcranial and the adaptative shift to modern humans. In *The Mousterian legacy: Human Biocultural Change in the Upper Pleistocene*, Trinkaus, E. (ed.). BAR International Series, Oxford.

Upledger, J.E. and Vredevoogd, J.D. (1983). *Thérapie cranio-sacrée*. IPCO, Paris.

Vaufrey, R. (1955). *Préhistoire de l'Afrique. I. Le Maghreb*. Publications de l'Institut des hautes études de Tunis, Tunis.

Vekua, A., Lordkipanidze, D., Rightmire, G.P., Agusti, J., Ferring, R., Maisuradze, G., Mouskhelishvili, A., Nioradze, M., Ponce de Leon, M., Tappen, M., Tvalchrelidze, M., Zollikofer, C. (2002). A new skull of early *Homo* from Dmanisi, Georgia. *Science*, 297, 859.

Vislobokova, I.A. (1990). The fossil deer of Eurasia. *Transactions of Paleontological Institute*, 240, 208.

Walker, A.C. and Pickford, M. (1983). New postcranial fossils of *Proconsul africanus* and *Proconsul nyanzae*. In *New Interpretation of Ape and Human Ancestry*, Ciochon, R.L., Corrucini, R.S. (eds). Plenum Press, New York.

Walker, A.C., Teaford, M.F., Martin, L., Andrews, P. (1993). A new species of *Proconsul* from the early Miocene of Rusinga/Mfangano Islands, Kenya. *J. Hum. Evol.*, 25, 43–56.

Wanpo, H., Ciochon, R., Gu, Y., Larick, R., Fang, Q., Schwarcz, H., Yonge, C., de Vos, A., Rink, W. (1995). Early *Homo* and associated artefacts from Asia. *Nature*, 378, 275–278.

White, T.D. and Harris, J.M. (1977). Suid evolution and correlation of African Hominid localities. *Science*, 198, 13–21.

White, T.D., Suwa, G., Asfaw, B. (1994). *Australopithecus ramidus*, a new species of Hominid from Aramis, Ethiopia. *Nature*, 371, 306–312.

WoldeGabriel, G., White, T.D., Suwa, G., Renne, P., de Heinzelin, J., Hart, W.K., Helken, G. (1994). Ecological and temporal placement of early Pliocene Homi-nids at Aramis, Ethiopia. *Nature*, 371, 330–333.

WoldeGabriel, G., Haile-Selassié, Y., Renne, P., Hart, W.K., Ambrose, S.H., Asafw, B., Heiken, G., White, T. (2001). Geology and palaeontology of the Late Miocene Middle Awash Valley, Afar rift, Ethiopia. *Nature*, 412, 175–178.

Wood, B.A. (1992). Early Hominid species and speciation. *J. Hum. Evol.*, 22, 351–365.

Young, C.C. (1932). On the Artiodactyla from the *Sinanthropus* site at Choukoutien. *Palaeont. Sinica Péking (C)*, 8(2).

Zouhri, S. and Aouraghe, H. (1997). Présence d'un *Equus* caballin et d'une forme asinienne dans les sites de Sidi Bouknadel (Rabat, Maroc) et de Mugharet El Aliya (Tanger, Maroc). In *14ème Colloque des bassins sédimentaires marocains*. Université Ibn Tofaïl, Faculté des Sciences, Kénitra, 24–27 September.

Index

C

D, E

F

G

H

K

L

M

N

O

P

Other titles from

ISTE

in

Biology and Biomedical Engineering

2020

CAZEAU Cyrille
Foot Surgery Viewed Through the Prism of Comparative Anatomy: From Normal to Useful
(Comparative Anatomy and Posture of Animal and Human – Volume 1)

DUJON Bernard, PELLETIER Georges
Trajectories of Genetics

2019

BRAND Gérard
Discovering Odors

BUIS Roger
Biology and Mathematics: History and Challenges

2016

CLERC Maureen, BOUGRAIN Laurent, LOTTE Fabien
Brain-Computer Interfaces 1: Foundations and Methods
Brain-Computer Interfaces 2: Technology and Applications

FURGER Christophe
Live Cell Assays: From Research to Health and Regulatory Applications

2015

CLARYSSE Patrick, FRIBOULET Denis
Multi-modality Cardiac Imaging: Processing and Analysis

2014

CHEZE Laurence
Kinematic Analysis of Human Movement

CLARYSSE Patrick, FRIBOULET Denis
Multi-modality Cardiac Imaging: Processing and Analysis

DAO Tien Tuan, HO BA THO Marie-Christine
Biomechanics of the Musculoskeletal: Modeling of Data Uncertainty and Knowledge

FANET Hervé
Medical Imaging Based on Magnetic Fields and Ultrasounds

FARINAS DEL CERRO Luis, INOUE Katsumi
Logical Modeling of Biological Systems

MIGONNEY Véronique
Biomaterials

TEBBANI Sihem, LOPES Filipa, FILALI Rayen, DUMUR Didier, PAREAU Dominique
CO_2 Biofixation by Microalgae: Modeling, Estimation and Control

Printed and bound by CPI Group (UK) Ltd, Croydon, CR0 4YY

15/09/2026

14962359-0005